Pink Collar Blues

Pink Collar Blues

WORK, GENDER & TECHNOLOGY

Edited by

BELINDA PROBERT
BRUCE W. WILSON

Melbourne University Press
1993

First published 1993
Design by Lauren Statham, Alice Graphics
Typeset in 10½/12½ point Baskerville
by Syarikat Seng Teik Sdn. Bhd., Malaysia
Printed in Malaysia by SRM Production Services Sdn. Bhd.
for Melbourne University Press, Carlton, Victoria 3053
U.S.A. and Canada: International Specialized Book Services, Inc.,
5804 N.E. Hassalo Street, Portland, Oregon 97213-3644
United Kingdom and Europe: University College London Press,
Gower Street, London WC1E 6BT

National Library of Australia Cataloguing-in-Publication entry

Pink collar blues: work, gender & technology.

Bibliography.
Includes index.
ISBN 0 522 84520 7.

1. Women—Employment—Effect of technological innovations on.
2. Sex role in the work environment. 3. Technological innovations—
Social aspects. 4. Office practice—Automation—Social aspects.
I. Probert, Belinda. II. Wilson, Bruce, 1951– .

331.4

Cover photograph: University of Melbourne collection,
University of Melbourne Archives.

Contents

Notes on contributors

Eileen Appelbaum recently jointed the Economic Policy Institute in Washington, DC as Associate Research Director. She has written extensively on the labour market experiences of women, including the effects of technology on women's jobs and the reasons for the expansion of part-time and contingent work arrangements in the United States. She is the author of, among other publications, the book *Back to Work* (1981), which examines the labour force experiences of mature women returning to work, and co-editor of *Labor Market Adjustments to Structural Change and Technological Progress* (1990). In addition Eileen has acted as consultant to the Office of Technology Assessment of the U.S. Congress on several volumes, including *Programmable Automation Technologies in Manufacturing* (1985) and *Trade in Services* (1988).

Suzanne Franzway is currently undertaking research on gender and power in Australian trade unions, extending her long-standing interest in feminist politics and practices. She is co-author of *Staking a Claim—Feminism, Bureaucracy and the State* (1989). Suzanne is a long-term, active member of unions including her academic union, and of the Women's Standing Committee of the South Australian United Trades and Labour Council, as well as the South Australian Working Women's Centre. She is Senior Lecturer in Sociology and Women's

Studies in the Faculty of Social Sciences, University of South Australia, and the Director of the Centre for Gender Studies.

Mike Hales works both theoretically and practically in the area of user-centred design, with a particular interest in the approach known as Computer Supported Co-operative Work. His previous work includes *Living Thinkwork—Where do Labour Processes Come From?* (1980) and *Science or Society: the Politics of the Work of Scientists* (1982). A 'turncoat engineer' (first degree in chemical engineering and operational research), he has worked for the past twenty years within critical theory and libertarian-socialist-feminist traditions, mostly in non-academic institutions. He is currently Senior Research Fellow in the Centre for Business Research at the University of Brighton Business School, England.

Miriam Henry is currently involved in research on the implementation of gender equity policies and on the new training and vocational education agenda, as part of a broader interest in the politics of education restructuring. She is a Senior Lecturer in the School of Cultural and Policy Studies at the Queensland University of Technology, where she teaches and writes on education policy, the sociology of education and women's studies. She has been active in her academic union, particularly in policy areas affecting women.

Cate Poynton teaches, researches and writes on a wide range of topics broadly concerned with language and social relations, with a particular interest in language and gender. Her work brings together theoretical and analytical perspectives from linguistics, cultural studies, feminist and post-structuralist theory. Cate is the author of *Language and Gender: Making the Difference* (1985), and various papers on interpersonal aspects of language, address terms and practices, language and the media, and children's writing. She is currently a Senior Lecturer in the Faculty of Humanities and Social Sciences at the University of Western Sydney, Nepean, where she teaches linguistics and women's studies.

Belinda Probert has published work on a variety of aspects of technological change and work reorganisation, including computer-based working at home (with Judy Wajcman), the establishment of remote offices, and the restructuring of clerical industrial awards. She is the author of *Working Life: Arguments about Work in Australian Society* (1989), as well as an earlier book on Protestant politics in Northern Ireland. Belinda is currently Senior Research Fellow at the Centre for International Research on Communication and Information Technologies (CIRCIT), in Melbourne, where she has been responsible for developing the Centre's research programme in the area of technological change, work and employment.

Judy Wajcman is currently doing comparative research on pay equity in Britain and Australia, as Principal Research Fellow in the Industrial Relations Research Unit, University of Warwick, on leave from the University of New South Wales, where she is Associate Professor in the School of Sociology. Her previous research has focused particularly on feminist approaches to science and technology, and on gender relations in the workplace. She is author of *Women in Control* (1983), *Feminism Confronts Technology* (1991) and co-editor of *The Social Shaping of Technology* (1985).

Juliet Webster has been researching in the area of gender and technology for well over a decade. She specialises in the study of women's office work, and is the author of *Office Automation: the Labour Process and Women's Work in Britain* (1990). Juliet is currently working on a second book, provisionally entitled *Gender and Technology at Work* (forthcoming), which will provide an overview of the research and debates in this area. She is Senior Lecturer in the Department of Innovation Studies at the University of East London.

Bruce Wilson is currently Director of the Union Research Centre on Organisation and Technology, which is a research intitiative of the Public Sector Union and the Australian Taxation Office. The centre is breaking new ground by collaborating with clerical workers in the conduct of action-oriented research. A key focus of the work is ensuring that the design of work and technical systems is both democratic and enhances the quality of work, as well as meeting organisational outcomes.

Bruce is seconded from the Department of Social and Educational Studies in the Institute of Education at the University of Melbourne, where he was the Director of the the Youth Research Centre. He is co-author of two books, *Confronting School and Work* (1984) and *Shaping Futures* (1987), and co-editor of *For Your Own Good: Young People and State Intervention in Australia* (1992).

Preface

This book is the outcome of a programme of seminars, workshops and discussions around the general topic of gender and technological change in the workplace, conducted during 1992. The programme brought together scholars and trade unionists, policy developers and action researchers with a common interest in both analysing the issues involved, and developing strategic responses of benefit to women workers. The impetus behind this programme came from two particularly ambitious Australian attempts to promote the reorganisation of routine white-collar work around better jobs and careers with greatly increased levels of skill—the restructuring of the main clerical industrial awards, and the transformation of work organisation in the Australian Taxation Office. As in all comparable countries, the lower status office workforce in Australia is overwhelmingly female, and any attempt to address the low status, low pay and lack of career opportunities for clerical workers immediately raises a series of questions about the significance of gender relations—in the historical construction of the occupations, the shaping of office technologies and the manner of their introduction, as well as the ability of the workers themselves to define and defend their interests.

The programme of events was designed to raise awareness of these issues and promote innovative analysis and action. We decided to involve a number of scholars and researchers from Australia, Great Britain and the United States who were able to make a major contribution to our understanding of one or other of these issues. We did not expect to find straightforward analogies for Australia from the experiences of women workers in different countries. On the contrary, as Eileen Appelbaum argues in her contribution, the introduction of information technologies into the office will have very different outcomes in different countries (and even within the same country) in terms of job opportunities, skill levels and career paths, depending on a range of institutional interests and constraints, and firms' strategic considerations. It is, however, by developing an international and comparative approach that we can more clearly identify not only the threats to women's employment entailed by rapid technological change in the workplace, but also strategic opportunities. Mike Hales's work, for example, raises tantalising questions about the scope for transforming low status women office workers into creative and skilled information systems designers and users in contexts beyond British local government.

In publishing the contributions which formed the basis of the Gender, Technology and Work programme we have two primary objectives. The first is to bring together in one volume a coherent set of articles which advance the theoretical understanding of the interaction between gender relations, work and technological change. The book begins with an overview, from an Australian perspective, of women's circumstances in the labour market, and in office workplaces particularly. It concludes with a discussion of the importance of trade unions for women workers seeking to turn technological change to their advantage. The second primary objective of the book is to promote further research, debate and action on the implications of gendered workplaces for social and technological change. We hope also that the book will help to link new networks of researchers as we are compiling a register of individuals and research centres engaged in relevant activities. Readers can pursue their own interests with the use of this register by making contact with us.

The Centre for International Research on Communication and Information Technologies (CIRCIT) and the Union Research Centre on Organisation and Technology (URCOT) have related interests in the relationship between organisational and technological change, and in the implications of these changes for social relations in workplaces. Given the particular significance of changes in organisation and technology for women office workers, the two Centres readily agreed to support the Gender, Technology and Work Programme.

That the programme has been organised and the book produced in a relatively short space of time is remarkable. It is a tribute to the energy, commitment and co-operative spirit of each of our contributors that this has been possible. Our own enthusiasm for the project has been fuelled by the response to the series: people from trade unions, various levels of government, educational institutions and community organisations, and from many parts of Australia, have all shown enormous interest in the project. We have greatly appreciated the participation of both contributors and their audiences.

Finally, the book owes much to the commitment and skills of Karen Davies, the URCOT Administrative Officer. Indeed, the book is the outcome of the efforts of many people, and we are grateful to all of them.

Belinda Probert and Bruce Wilson

CIRCIT/URCOT
1/4, Riverside Quay
South Melbourne
Victoria 3205

1

Gendered work

BELINDA PROBERT
BRUCE W. WILSON

It comes as a surprise to many people that women workers are enthusiastic about the introduction of new technologies into the office. They see new technology as making their jobs more interesting and substantially increasing their skills, as well as enhancing their efficiency (Liff 1990, pp. 45–8). Women office workers also express a strong interest in undergoing further training to develop their computer skills, even though it is widely recognised that such training will not lead to pay increases or promotion (Probert 1992).

This is particularly significant as clerical work is the single largest source of employment for women in most advanced industrial societies. In Australia over one million women are categorised as clerical employees. Clerical jobs provide 32 per cent of women's jobs, which is almost twice as much as any other occupational grouping. In the United Kingdom an even larger proportion of women workers are employed in clerical occupations, while in the United States four out of five of all clerical workers are women.

During the 1980s there has been massive investment in information technologies affecting office workers particularly. In the 1950s, information technology in the United States represented 2.5 per cent of all industry's capital stock; this figure rose to 5.8 per cent in the 1970s and 12.5 per cent in 1985. In the banking industry, however, the information technology share of capital stock rose from 1.7 per cent in the 1950s to

3.9 per cent in the 1970s, and to 26.3 per cent in 1985. In all service sectors, the corresponding figures were 3.3 per cent (1950s), 7.7 per cent (1970s) and 15.5 per cent in 1985 (OECD 1988, p. 52).

In the mid-1970s Harry Braverman painted a gloomy picture of the future of office work, in which he saw the same pressures towards job fragmentation and deskilling as those which had created the great assembly lines of mass production in manufacturing. By the mid-1980s, however, many observers of office work were adopting a rather different view, and pointing to the emerging evidence about job satisfaction. Many writers have argued that computer technologies represent a radical break from previous forms of technological change in the workplace. Shoshana Zuboff, for example, has argued that there is a 'crucial distinction' between computer systems and the office machines that preceded them.

> The further development of office machinery begets only new and more specialised office machinery. Typewriters, adding machines, filing systems—equipment may become more sophisticated, but its function does not change. In contrast, the further development of computer systems unleashes their informating potential (1988, p. 172).

For Zuboff the critical factor distinguishing information technology from earlier generations of machine technology is its ability 'to create a vast overview of an organisation's operations, with many levels of data coordinated and accessible for a variety of analytical efforts' (p. 9). For the full potential of the technology to be realised all employees need to obtain the skills necessary to carry out a wide range of tasks using their access to the organisation's data base.

Much of the literature on office automation has adopted a similarly optimistic view of the possibilities for work organisation, job design and skill levels. These are seen as strategic issues rather than matters determined by the technology itself, and the optimism derives from the argument that computer technologies are best exploited by workers with increased skills and autonomy. For Zuboff and many other writers the decisive factor in determining how new office technologies are implemented is management's willingness to let go of traditional hierarchies and to give up its monopoly over information.

Optimism about the beneficial effects of computerisation has not been confined to the white-collar workplace. On the contrary, it has been most strongly expressed in recent debates about the significance of 'new production concepts' in core manufacturing industries, which no longer feature the Taylorist and Fordist ideas about fragmenting tasks and divorcing conception from execution (Piore & Sabel 1984; Kern & Schumann 1987). According to authors such as Harvey (1989) changes in the nature of global markets have reduced the competitive advantages of the mass production model based on a standardised product at the lowest cost. Success in these new markets will go to firms able to respond flexibly to a wide range of ever-changing 'market niches', with increasing emphasis on quality rather than price competition. What makes this kind of flexibility possible is, in turn, micro-electronic circuitry which 'has progresively eroded . . . the traditional distinctions between mass and specialist production' (Sorge & Streek 1988, p. 30).

In Australia, the case for the beneficial effects of computer based workplace technologies has been most strongly put by John Mathews (1989) who has sought to persuade the trade union movement that they offer the opportunity to establish more rewarding and less hierarchical work practices and more democratic social processes. For Mathews the new economic conditions require a highly skilled and motivated workforce to enable firms to respond to demands of the new world market for more specialised products, rapid innovation, and for quality rather than quantity in production.

There has been widespread debate about the validity of the theoretical analysis which underpins these claims about the transition from Fordism to post-Fordism, or from mass production to flexible specialisation. Empirical evidence about the effect of new production technologies does not, however, support the notion of a technologically determined trend towards more highly and flexibly skilled workers. Comparative studies of how similar technologies are used in different countries reveal contrasting outcomes, reflecting the importance of the way firms see their existing organisational and skill strengths and the broader industrial relations framework. Managers in the United States continue to show very little interest in training and skill development, compared to Germany for example (see Appelbaum, in this volume).

While national variations in social institutions and historical skill levels are key factors which affect the extent to which new production concepts are adopted, structured patterns of difference exist also *within* nations. In particular, much of the discussion about historical trends has been gender-blind, and has failed to ask 'whether women and men are located in the labour force in such a way that they will benefit similarly from any new emphases on skilled labour' (Jenson 1989, p. 141).

> If, for example, employers make use of the female and male labour force in different ways, if the development of a more 'flexible' labour force also means rising rates of feminisation, if only men are likely to be 'flexible specialists', then a world of post-Fordist flexible specialisation is very different—and less benign—than that which Piore and Sabel assume (pp. 144–5).

Even where some progress has been made in addressing labour market segmentation, as in Sweden, the dynamics of gender relations continue to be significant in shaping the implementation of new technologies.

Trade unions and new technology

The sheer scale and pace of technological innovation over the last twenty years has forced many unions to develop a more strategic response than simply trying to protect jobs (or job holders, rather). In some countries this has taken the form of seeking to win generalised rights for unions in the process of technological change, such as the Swedish codetermination movement (Gustafsson 1986). In others the focus has been on negotiating new technology agreements with individual employers under which unions are able to influence the manner in which new technologies are introduced. For the most part, these initiatives have left unions and employees in the position of reacting to the plans of employers. Comparing the legal provisions in France and Germany and the content of new technology agreements, Michele Tallard concludes that involvement is 'often limited to discussing the *consequences* of NT; scarcely ever are unions involved in investment decisions or the choice of technology. Employers resist this as an infringement of their prerogatives, and sometimes unions themselves do not see this as their function' (1988, p. 293).

There have, however, been isolated instances where unions have been able to challenge the design and selection of new workplace technologies, and these have been the subject of widespread international interest. Perhaps the best known of these is still the Swedish project called UTOPIA, which looks at training, technology and products from a quality of work perspective. The project, developed by the Swedish Centre for Working Life, was set up to give print workers the resources to intervene collectively in the design of new computerised technology in ways that enhanced their existing skills. Inspired by this kind of example, researchers within and outside the trade union movement have argued that trade unions need to develop their own vision of the design and use of particular technologies in particular workplaces. As Murray puts it:

> Intervention in the design stages of computer-based systems would be a crucial aspect of such a policy, which would recognise technological change as integral to a larger restructuring of many workplaces and would combine existing and well-established trade union policies and practices with new initiatives (1989, p. 520).

In Australia, the trade union movement response to the restructuring of employment around new production technologies has centred on two broad strategies. First, individual unions have attempted to secure specific new technology agreements, one or two of which have succeeded in challenging the employer's control over key elements of the selection and implementation of new technologies. The most commonly cited of these is perhaps the agreement which came after a major dispute between Telecom and the telecommunications engineers in 1978, over the introduction of computer-controlled switching apparatus (Palmer 1988). The engineers were not opposed to the new technology itself, but to the job design proposals that accompanied it which would have created an elite corps of maintenance staff. The outcome of the dispute was essentially a compromise between management's proposal and the union's preferred form of work organisation, but it also led to the negotiation of a general technological change agreement in 1980. Following on from several specific industrial struggles over new technology, the Australian Council of Trade Unions successfully campaigned for employees to be notified and consulted by employers about proposed technological changes. However, these rights come into force only after management has decided on the changes to be made (Davis & Lansbury 1989, p. 107). Furthermore, the ability of unions to exploit these provisions in meaningful ways is heavily constrained by their lack of resources, both financial and technical.

Initiatives by unions to increase their influence over technological change in Australia have been more successful in the public than in the private sector. In 1990, a path-breaking agreement was reached between the Australian Taxation Office (ATO) and the Public Sector Union (PSU) governing union participation in a massive programme of workplace computerisation. In this case the union sought to establish not only guarantees about job protection, work organisation, occupational health and safety, job redesign and career development, but also campaigned for a clear process of union participation at all stages and all levels of the change process, known as 'modernisation'. Inspired by the Scandinavian projects in which unions had successfully won a role in the design and selection of new technologies, the PSU specifically included a claim for resources that would allow them to establish an independent advisory unit to carry out research on their behalf and provide advice on modernisation. A central task for this unit was to enable the PSU to challenge the ATO at the level of systems design and selection, not merely at the point of introduction. Flowing from this agreement, the Union Research Centre on Office (now Organisation and) Technology was established in 1991, and in 1992, it co-sponsored the 'Gender, Technology and Work Programme' which is the basis of this book.

As well as these somewhat isolated examples of new technology agreements, the trade union movement has participated in a general strategy of updating all industrial awards around structures based on skill and career development. Recognising the need to move away from traditional craft boundaries and notions of job ownership, the union movement embraced the principles of multi-skilling and flexibility. In 1987, the national industrial relations court required employers and unions to pursue 'structural efficiencies' as a basis for future wages increases. In the case of the new Metal Industry Award (the blueprint for award restructuring), narrowly defined trades skills have been broad-banded to include the performance of broader, more generic tasks. Over 300 traditional classifications have been reclassified into fourteen broad bands of skill level.

Award restructuring has been a painstaking, complex and tense process, challenging as it does the privileges of craft distinctions and traditions, some of which can be traced to medieval times. The process has extended well beyond manufacturing to encompass all industry sectors. Different issues and agreements have emerged in the restructuring of major clerical awards, such as the Commercial Clerks Award (covering private sector clerical workers), the Commonwealth Bank Officers Award, and the agreements covering the Australian Public Service. In many circumstances, clerical workers have been 'multi-skilled' for decades, but their skills devalued because they have been performed largely by women. The benefits to employers from acknowledging and rewarding the extent and value of multi-skilling, in the present context, are limited, to say the least.

In the Australian Public Service, restructuring of the terms and conditions of clerical workers was linked to a process of participative job design, resulting in a single, multi-skilled career path for all clerical workers. Ironically, this removed from some women their technical expertise, specifically keyboarding, and undermined their sense of status and worth in the organisation. Circumstances such as these have been justified on the basis that the overall prospects for skill development and promotion for women workers have been enhanced.

Gender at work

If technological change in the workplace can be used to enhance the skills of workers, are men and women workers in a similar position to benefit from this? As women entered the paid workforce in increasingly large numbers during the 1970s and early 1980s, feminist analysis tended to focus on the ways in which women's family responsibilities constrained them from participating on an equal footing with men. The concentration of women in a limited number of low status occupations was commonly linked to the unequal domestic division of labour and to other 'disabilities' which women brought with them into the labour market. Girls' socialisation and schooling in particular were identified as problems since women were (and still are) far less likely to have studied science subjects or to have engineering or computing qualifications, or to have done an apprenticeship in anything other than hairdressing.

During these years women's organisations lobbied strongly and successfully around these issues, campaigning for maternity leave, quality child-care provisions, and legislation prohibiting discrimination in appointments and promotions. There has also been a great deal of research in the education system, exploring a range of issues relating to girls' participation in science and technology programmes, and in traditional 'male' trades.

Much of the activity around equal opportunity and affirmative action campaigns involved an attempt to give women the chance to move out of low-skill, feminised ghettoes of employment, and to bring about a more equal distribution of occupations between the sexes. The barriers were defined primarily as either domestic or simply historical—women were confined to certain kinds of employment because of attitudes and domestic handicaps which were increasingly obsolete. The central human obstacles, from this perspective, were unreconstructed husbands and employers who benefited from a pool of labour which, they assumed, had no interest in training, promotion or job satisfaction.

This approach suggested some grounds for optimism about the impact of computer-based production systems on women workers. If employers were increasingly aware of the productivity advantages which could flow from improving the skills of their employees, both women and men could benefit, even if many women were starting from a less skilled and less well organised base.

By the late 1980s, however, many observers were doubtful about the adequacy of the equal opportunity approach and the emphasis on removing women's labour market handicaps. Labour market segmentation has remained deeply entrenched which, as Eileen Appelbaum argues in her chapter, has profound implications for women's ability to benefit from technological change. The research emphasis has shifted from 'women' to 'gender' (Beechey 1987) and to the continuities in relationships between women and men which keep the former subordinate to the latter. In their influential book *Gender at Work*, Anne Game and Rosemary Pringle suggested that the problem was not in fact external to the workplace, but that it was reproduced continuously within the workplace itself, in a remarkably flexible and adaptable manner. They argued that there was nothing inherent in any particular job which made it either appropriately female or male. Jobs could indeed change their sex-typing. The only thing which remained fixed was that there should be a distinction between men's and women's work. They argued that the distinction remains and is constantly reproduced because 'gender is not just about difference but about power' (Game & Pringle 1983, p. 16).

Since that book was published ten years ago, feminist scholars have done much to elucidate the basis of these power relations and the ways in which women's subordination is sustained within the workplace. Technology itself, and the associated social construction of skills, is a site of gender struggle. Cynthia Cockburn's work, for example, has been particularly concerned with the way in which technological competence is identified with masculinity. Her study of compositors (1983b) clearly illustrated how male workers responded to the threat of new technology not by resisting the technology but by ensuring that the right to use the technology was confined to a small group of men, organised within a powerful craft union. Similar patterns have been found in the clothing industry (Phillips & Taylor 1986).

Central to the efforts of these male craft workers to protect their privileged status and pay was their ability to defend their work as skilled, and to differentiate it from other kinds of work. As Phillips and Taylor have argued, 'skill has increasingly been defined *against* women—skilled work is work that women don't do' (1980, p. 63). Elsewhere, Phillips has indicated that the hierarchies of skill to be found in the workforce are not simply the outcome of capitalist production, but 'express at the same time a system of male dominance in craft identity which is inextricably (if confusingly) linked with masculinity . . . jobs are created as masculine and feminine, with their skill context continually re-drawn to assert the dominance of men' (1983, p. 102).

In assessing how women workers are likely to benefit from new computer-based production systems, it is clearly essential to question whether skill is likely to remain such a gendered concept. The male bias in the definition of skilled work has resulted typically in women's work being defined as unskilled, and hence of relatively limited value. In the present context of rapid technological change it is particularly important to challenge the basis on which women's skills have been valued historically if women are to benefit from the current trend towards explicit skill evaluation. The process by which gender bias is eliminated from job evaluation schemes is, however, far from straightforward. Women workers themselves tend to belittle the work they do by describing their skills as feminine attributes rather than as learned skills. In this volume Cate Poynton describes a particularly interesting project designed to help women clerical employees describe the skills they use in more technical language in order to ensure adequate recognition and remuneration. As Jane Jenson concludes,

> restructuring the labour process so as to privilege skilled work and workers will further marginalise women unless political actors challenge long-standing processes which isolate women from machinery and which define women's skills as talents. Carefully constructed strategies in pursuit of equal pay for work of equal value—which by their very nature politically question popular notions of skill and value—may be part of such a process (1989, p. 155).

It is tempting to think that the direction of industry restructuring itself, encouraging greater participation and decentralised decision-making within a framework designed to promote and reward the acquisition of skills, will lessen the tendency for skilled, *male* craft workers to exploit gender divisions in their own defence. The evidence about gender and power in the workplace suggests, however, that there has been an intrinsic connection with male domination in the domestic sphere. Male workers have been as motivated by their desire to maintain authority in the home, as they have to preserve industrial power and privilege. The threat to men's domestic authority has been growing steadily over the last twenty years, yet there is plenty of evidence that men continue to resist this challenge. Socially acceptable definitions of masculinity continue to hinge on the idea that it involves the ability to do things that women cannot do. Clare Burton, for example, explains the failure of many EEO strategies in this way, pointing to the fact that 'part of the satisfaction in some jobs derives from the belief that it is *masculine* work, that women could not perform it adequately' (1991, p. 6). This view, which has been called 'the fear of contagious effeminacy', manifests itself concretely in the phenomenon known as 'tipping' in the American literature—the tendency of men to leave an occupation as fast as women enter it. It is the continued relevance of processes such as these which help to explain the extraordinary tenacity of job segregation by sex, and the continued powerful association of many workplace technologies with masculinity.

It would be a mistake, however, to characterise this process of gender differentiation as simply one in which men either seek to prevent women playing at all, or insist on moving the goal posts continually. Women do not 'fail' to make headway in male occupations simply because of unequal opportunity and discrimination. Cynthia Cockburn's work has been particularly useful in helping to explain the limitations of approaches which focus on encouraging women to take up traditionally male subjects or occupations. Cockburn argues that 'we should look instead at the environments offered to women and how they may actively choose to absent themselves'. No amount of exhortation, encouragement and opportunity will work as long as occupations such as engineering involve 'masculine patterns of relationships . . . a camaraderie based on the exchange of anecdote and slander concerning women' (1983a, p. 18). The importance of this kind of approach was apparent in a project which drew on the ideas of Sherry Turkle to analyse women's participation in different computer science courses in universities in Norway, Denmark, Sweden and Finland. In explaining the very small numbers of women choosing to study computer science when it is located in technical departments, the researchers emphasised the importance of the culture of the computer-freaks and the dominance of this culture within that course. They concluded that the female students wished actively to distance themselves from that culture 'because the computer-freaks represent that which they do not want to be. The female students do not want an intimate relation with a computer. Intimacy belongs to *people* and not to machines' (Hapnes & Rasmussen 1991, p. 413).

The explanations for women's continued concentration in a narrow range of relatively low status occupations have become more complex as gender has been recognised as a central organising principle not only of family life but the workplace itself. At the same time, the effects of this extraordinarily persistent division have been found to be more far-reaching than many early feminist critiques suggested. Sex segregation has not merely perpetuated the low skill, low pay status of women workers, but it has had a profound impact on the shape of feminised jobs and the technologies introduced around them. As Sonia Liff has argued, for example, 'saying that office jobs are women's work means more than just drawing attention to the fact that a lot of women work there. It also highlights the forms of work organisation and relations which operate' (1990, p. 44). The fact that clerical workers have for most of this century been overwhelmingly female would seem to be a vital fact in explaining the significance of the personal service element in clerical labour. It is almost impossible to imagine a man in most secretarial jobs, not simply because of their low status, but because of the way they include many wifely or motherly tasks, from making cups of tea to reminding men about what needs to be done that day. 'We can at least state with some certainty that the entire personal service aspect of clerical labour (as in virtually all the service work performed by women) would not in this case have developed in this way' (Phillips & Taylor 1980, p. 62; also Pringle 1991).

These ideas have been developed in a number of ways, including the suggestion that organisational structure itself is not gender neutral. As Acker has argued, 'In organisational logic, both jobs and hierarchies are abstract categories that have no occupants, no human bodies, no gender' (1990, p. 49). However, in reality a job already contains 'the gender-based division of labour and the separation between the public and the private sphere. The concept of "a job" assumes a particular gendered organisation of domestic life and social production' (p. 149). 'There is no place within the disembodied job or the gender neutral organisation for other "bodied" processes, such as human reproduction or the free expression of emotions' (p. 151).

The potential of computer-based technologies to support substantial changes in dominant patterns of work organisation is a central element in contemporary debate. Shoshana Zuboff, (1988, chap. 8) for example, has written persuasively about the way computers should be used to flatten hierarchical structures and eliminate the wedge of middle managers who have historically been required to supervise the ranks of unskilled and unmotivated process workers. There is little evidence, however, that new computer technologies have any determining effect on patterns of work organisation and new job design. On the contrary, it appears to be the form of work organisation which influences implementation of the new technologies. Juliet Webster's research (1990a) into the introduction of word processing in a number of different clerical workplaces clearly illustrates this: where there was fragmented and routinised work this was organisationally driven, and it was a pattern which was reinforced by the introduction of word processors. By contrast, office workers who already had jobs with some autonomy, flexibility and variety found these characteristics enhanced by the addition of word processing technology (1990, chap. 3).

This is not to suggest that the outcome of office automation is simply contingent on another random variable—work organisation. Work organisation is itself a factor which is highly gendered. Firms and industries which are able to rely on a generous supply of low-cost female labour are likely to adopt and maintain forms of work organisation which are low-skill and low-pay. This tendency is in turn reinforced by employers' assumptions about the marginal attachment of women workers to the labour market. As Peter Albin and Eileen Appelbaum have observed in the American context, 'the long-standing reluctance of employers to provide costly training for female clerical workers has contributed to the managerial bias in favour of more algorithmic forms of computer rationalisation' (1988, p. 145). It could be argued that this is precisely the case in the Australian banking industry which, according to John Mathews, has behaved in 'typically Fordist fashion', proceeding 'down the Taylorist road, equating on-line access with deskilling, and assuming that counter staff would need less and less training as the terminal took over more and more of their functions' (1989, p. 62). What Mathews fails to address in his analysis is the role that gender plays in discouraging banks from choosing his preferred post-Fordist route. Indeed it is hard for him to explain this development except in terms of the inadequacies of bank management.

Eileen Appelbaum argues in Chapter 4 that gendered patterns of work organisation mean that low status female clerical workers are particularly likely to find their jobs eliminated altogether by the introduction of computer technologies. Where large amounts of routinised white collar work remain, computer and communication technologies are likely to be used to pursue further labour cost savings by, for example, relocating back-office clerical work offshore in lower wage countries such as Jamaica and Barbados. The conclusion is that women stand to fare badly as a result of the computerisation of work wherever technological change is not accompanied by work reorganisation and job redesign which has been informed by a gender analysis. Addressing these issues in a systematic way is likely to have far more profound implications for women's employment opportunities than strategies which focus on giving women equal opportunities for promotion within the existing job structure.

Theory into practice

Reviewing the ways in which the analysis of women's unequal experience of paid employment has developed over the last twenty years it is easy to feel disheartened. The entry of large numbers of women into the labour force has done little to alter the distinction between male and female jobs, or to improve the overall quality of working life. Some progress has been made in eliminating the most obvious factors contributing to unequal pay, but women are concentrated overwhelmingly in a very narrow range of relatively poorly paid occupations, such as clerical and sales.

Much of the recent feminist scholarship in this area has identified critical gender issues in the way technological change is implemented, but it has tended to treat the new technologies themselves as gender neutral. However, as Judy Wajcman's chapter suggests, technological objects themselves can be 'shaped by the operation of gender interests' (p. 35). Similarly, Juliet Webster's chapter in this book seeks to show that the evolution of word processing technologies was shaped by the assumption that typing was women's work.

Other recent research has attempted to identify the implications of the 'social shaping' perspective for new points of strategic intervention. Mike Hales, for example, in Chapter 6 of this book, takes up the question of the social interests which shape the design and implementation of new workplace technologies. The rhetoric of user-centred systems design and of user participation is one which has been so diluted by consultants selling systems packages that it bears little relation to the radical intentions of the Scandinavian experiments. However, even a rhetorical commitment to the idea that successful implementation depends on user co-operation provides opportunities for intervention. The Union Research Centre on Office (now Organisation and) Technology (URCOT) has been funded by the Australian Taxation Office (ATO) because the ATO recognises that the involvement of the Public Sector Union, and the expertise of its members, is crucial to the overall success of the Modernisation venture. URCOT has been able to develop a model of action research with union members, which builds on the Scandinavian tradition. Central to this approach has been the establishment of Investigative Work Groups, in which union members affected by particular change initiatives have volunteered to work collaboratively on research projects on issues of major concern. The outcomes of the research have informed the Union's negotiating positions, and have contributed directly to the change process.

A major challenge for researchers involved in the design and implementation of workplace computer systems is to recognise the gendered experiences and position of the 'users'. In a culture which systematically defines women as technologically incompetent, how are women of low employment status to contribute to a process with generally all-male technical specialists?

Significant experiments have been conducted in Britain and Scandinavia which provide concrete strategies. In particular there are examples of the use of workplace study circles within which women can talk about the work they do and analyse their own social and technical skills (Green et al. 1991). For women's organisational knowledge and tacit skills to be utilised rather than ignored—if they are to become what Mike Hales (1991) has called 'constructive end users'—then the process through which it is introduced needs as much attention as the technological choices themselves.

In the 'Human Centred Office Systems Project' (Green et al. 1991) the researchers negotiated an agreement with a large City Council and trade union branch to undertake a collaborative project to develop a new computerised library system. A central element of the agreement was the commitment to explore new forms of consultation or involvement for the 400 library assistants on clerical grades, almost all of whom were women. In this the researchers felt they were extremely successful, implementing a series of workplace study circles which met in work time, for an average of seven half-days each. They concluded that the circles succeeded in 'demonstrating that opportunities existed for women library assistants to play an active role in the planning of a new computerised system' (p. 224). The circles also prompted the setting up of a new structure—the mixed design team with members drawn from all levels of the library staff—to take the development process forward.

While it is important to draw attention to individual case studies in which gender has been recognised as a fundamental factor shaping the process of technological change in the workplace, it is also necessary to recognise the broader environment and its impact on the spread of such workplace initiatives. It is ironic that American writers should be providing such interesting theorising about gender and systems design when their own economic and industrial relations environment is so hostile (see chap. 4). Within Australia the centralised and highly regulated industrial relations framework has provided a number of positive opportunities over the last ten years. Sheltered by a Labor government committed to increased industrial democracy, the new skill-based award structures have given women an opportunity to campaign for the proper recognition of women's skills, and the emphasis on creating career paths has created new opportunities in industries traditionally full of dead-end jobs such as the hospitality industry.

In all the enthusiasm for gender-aware skill evaluations there has, however, been a tendency to imply that women's skills have been ignored, undervalued and unrewarded because of some oversight in the past—an oversight which everyone is now happy to correct. Yet this is to lose sight of the most fundamental feminist insights into the structure and reproduction of gender relations—that they are relations of power and domination from which both men and employers have stood to gain. The vested interests which still stand opposed to the re-evaluation of women's work can be seen, for example, in the case of clerical work (Probert 1992). The restructuring of the Victorian Commercial Clerks award provided an opportunity to identify different levels of word processing skills and to differentiate these from typing. These changes were resisted strongly by the employers' representatives, for reasons which are directly related to the feminised nature of clerical jobs and employment. In other areas of work it has been argued that both employers and employees stand to gain from award restructuring. In the metal trades there were obvious benefits to employers from the new skill classifications since they created a far more flexible workforce, and addressed real shortages of skills. In the clerical area there are no obvious benefits to employers in the new award structure, mainly because many clerical employees are already multi-skilled and the change simply means having to pay more for those skills.

Just as award restructuring still presents real opportunities to improve women's employment experiences in Australia, so other trends in the contemporary industrial relations agenda present very real dangers for women. In particular, the move away from a centralised wage fixing system to one with a primary emphasis on enterprise bargaining, and productivity-linked pay rises, poses serious threats. Women tend to work in the service industries where productivity is far harder to measure; they are much more likely to work in small firms which are difficult to unionise and where opportunities for career development are limited; and they are likely to be in a weaker position than men to enforce their interests in enterprise level bargaining.

The continued role of trade unions in the new post-Fordist workplace is, of course, critical to the outcome of these changes for employees. In this volume Suzanne Franzway and Miriam Henry take up the question of how far Australian trade unions are capable of transforming their industrial agendas to reflect the interests of women workers. As elsewhere, unions remain dominated by men, even in industries dominated by female employment. An awareness of the absolute necessity of unionising the growing female workforce if the union movement is not to fade into insignificance has galvanised some thought on the matter (Berry & Kitchener 1989).

In a review of the recent history of British trade unions' success with New Technology Agreements, Fergus Murray (1989) argues that they have failed to recognise the central importance of gender relations as an organising principle of the workplace. 'As such, trade union technology bargaining strategies are implicated in the reproduction and exacerbation of workplace and labour market divisions which tend to work systematically against women employees and women trade union members' (p. 518). Looking at one particular union in the banking industry, Murray shows how it has pursued two parallel and separate policies 'which could be mutually reinforcing if linked together: one, technology policy, attempts to negotiate the introduction of new equipment and computer systems; while the other, equal opportunities policy, aims to reduce and eliminate formal and informal discrimination against women employees' (p. 524). According to Murray, failure to recognise women's specific experience of and concern with massive technological change meant that the technology agreement could constrain women's employment opportunities even further by legitimising the bank's attempt to formalise a two-tier recruitment strategy.

By contrast, the Swedish Bank Employees Union has found ways of integrating the two strategies with more fruitful outcomes. Equal opportunity activists see the possibility of using technical change as an opportunity to challenge existing work organisation to reverse the trend towards a polarised job structure. 'In particular the union is attempting to promote the creation of multifaceted, multi-skilled jobs through the enlargement of clerical jobs into areas of traditionally male work such as loans, securities and foreign currency transactions' (p. 529). If the gender division of jobs is to be addressed then unions must ensure that new technology agreements are applied together with equal opportunity policies in the areas of work organisation, job design and training.

One of the most interesting elements in this Swedish example is the significance not only of differently formulated strategies, but the use of different kinds of processes for mobilising members. In particular, the union has relied on the formation of workplace study circles as the means to encouraging the participation of the end-users of new technologies. The importance of decentralised and informal organisational forms for the meaningful participation of women workers and trade unionists has been confirmed by a variety of international feminist researchers. These kinds of study circles succeed in mobilising women, according to Murray, 'because they provide an independent organisational base responsive to women's concerns within waged work and the home' (p. 532). The question then becomes one of how their insights and initiative become part of a broader programme of industrial and political action. Ultimately, these issues can only be resolved within the larger industrial and political environment.

This is, and always has been, a major challenge for women workers. Insofar as men's and women's interests conflict, trade unions have invariably failed to pursue the concerns of women workers, whether related to technological change or otherwise, to their ultimate conclusion. Such a strategy, if pursued, would threaten in one form or other the structural relations and power on which the very masculinity of many men is based.

The contributions to this book address this structural dilemma from a variety of perspectives. The theme which persists through these chapters is that women workers should not underestimate the scale and complexity of the task of realising the benefits of technological change; yet, at the same time, there are significant strategies emerging which, if informed by a gender analysis, offer real prospects of decent work for many women.

2

The masculine mystique

A feminist analysis of science and technology

JUDY WAJCMAN

In a book about technological change and paid work, it is appropriate to provide a broad overview of the theoretical debates that have, to a greater or lesser extent, informed more specific studies.[1] Over the last fifteen years an exciting new field of inquiry has emerged, concerned to develop a feminist perspective on science and technology. It has disparate intellectual and political roots and is by no means confined to analyses of workplace technologies. Its objective has been to show that science and technology are shaped by gender relations. Only by bringing a rigorous theoretical perspective to bear on empirical research can we see through the masculine mystique attached to technological activity, and understand the ways in which gender inequality is produced and reproduced at work.

Much of the recent sociological work on technology has been engaged in developing a critique of technological determinism (MacKenzie & Wajcman 1985). This is the theory that technology is an independent, autonomous factor causing fundamental changes in society. Although a critique of technological determinism may seem old hat, this is the single most influential theory of the relationship between technology and society. Although it has been common among some sociologists to acknowledge that technology is socially determined and to criticise theories of post-industrial society in these terms, theorists such as Alvin Toffler (1980), Andre Gorz (1982) and Barry Jones (1982) have given us new versions of the same old positions. In my view, many protagonists of post-modernism are in fact drawing upon the old post-industrial theorists and reworking the same familiar themes about how changes in technology, in this case the new information and communication technologies, are causing a new form of society and consciousness to emerge. There is also an equally fashionable and related debate about post-Fordism and 'flexible specialisation' which gives primacy to technology in bringing about a fundamental reorganisation of production. As a result of this approach, most sociologists have tended to concentrate on the 'effects' of technology and on the 'impact' of technological change on society. This is a perfectly valid concern, but it begs a crucial question. What has shaped the technology that is having effects? What has caused and is causing the technological changes whose impact we are experiencing?

Since the mid-1970s, feminist researchers and activists have also been examining the effects of automation, looking particularly at women's employment. The introduction of computer-based technologies into offices is the focus of one strand of this research, mainly because the majority of workers in clerical and secretarial occupations are women. (It is also the case that these groups are being disproportionately affected, as the office is the prime site of technologically induced change.) This research forms the basis for many of the generalisations about women's work experience. What has received less attention is the way different interests have shaped the technologies of the workplace.

It is these questions that the 'social shaping' approach to technology has sought to address. First, it concentrates on opening up the 'black box' and unpacking the processes of technological development and implementation. The point of this is to add to our understanding of the technical factors that shape technologies a recognition that there are social, economic, political, ideological and organisational influences at work, creating points of choice and negotiation within the process of technical change. Second, the approach recognises that the relationship between society and technology is not a linear one between two separate elements, but is reciprocal. The newly emerging sociology of technology looks at the effects of society on technology as well as the effects of technology on society.

To date, however, studies of technology have been deficient in their treatment of gender issues. Although there is now a burgeoning feminist literature in this area, exploring women's relationship to, and experience of, technology, it is very segmented. So there are disparate publications dealing with, for instance, women and reproductive technology, domestic technology, or women and workplace technology. There has been little attempt to develop a comprehensive theory of the extent to which gender relations shape technology. Furthermore, just as the sociological literature has largely ignored gender issues, the new feminist debate often takes little account of all the radical science and sociology of technology debates and is carried out as if such antecedents did not exist.

This chapter seeks to bring together the theoretical insights from these different sources to help illuminate the relationship between gender and workplace technologies, the theme of this book. I begin by presenting an overview of feminist approaches to science and then go on to look at technology. The chapter starts at gender and science because this subject has attracted much more theoretical debate than the related subject of gender and technology. It will become apparent in what follows, however, that feminists pursued similar lines of argument when they turned their attention from science to technology. In the final section I illustrate my approach with some examples of technologies from home and work.

The sexual politics of science

Interest in gender and science arose out of the contemporary women's movement and a general concern for women's position in the professions. Since the early 1970s, the publication of biographical studies of great women scientists has served as a useful corrective to mainstream histories of science in demonstrating that women have in fact made important contributions to scientific endeavour. The biographies of Rosalind Franklin and Barbara McClintock, by Anne Sayre (1975) and Evelyn Fox Keller (1983) respectively, are probably the best known examples. Recovering the history of women's achievements has now become an integral part of feminist scholarship in a wide range of disciplines. However, as the extent of women's exclusion from science became more apparent, the approach gradually shifted from looking at exceptional women to examining the general patterns of women's participation.

There is now considerable evidence of the ways in which women have achieved only limited access to scientific institutions, and of the current status of women within the scientific profession. Many studies have identified the structural barriers to women's participation, looking at sex discrimination in employment and the kind of socialisation and education that girls receive which have channelled them away from studying mathematics and science. Explaining the under-representation of women in science education, laboratories and scientific publications, this research correctly criticises the character and construction of feminine identity and behaviour encouraged by our culture.

However these authors mainly pose the solution in terms of getting more women to enter science—seeing the issue as one of equal access to education and employment. Rather than questioning science itself, such studies assume that science is a noble profession and that if girls were given the right opportunities and encouragement they would gladly become scientists. It follows that the current deficiency is seen as a problem that could be overcome by a combination of different socialisation processes and equal opportunity policies.

This approach locates the problem in women (their socialisation, their aspirations and values) and does not ask the broader questions of whether and in what way science and its institutions could be reshaped to accommodate women. The equal opportunity recommendations, moreover, ask women to exchange major aspects of their gender identity for a masculine version without prescribing a similar 'degendering' process for men. For example, the current career structure for a professional scientist dictates long unbroken periods of intensive study and research which simply do not allow for child-care and domestic responsibilities. In order to succeed women would have to model themselves on men who have traditionally avoided such commitments. The equal opportunities strategy has had limited success precisely because it fails to challenge the division of labour by gender in the wider society.

When feminists first turned their attention to science itself, the problem was conceived as one of the uses and abuses to which science has been put by men. For example, feminists have highlighted the way in which biology has been used to make a powerful case for biologically determined sex roles. They set out to demonstrate that biological inquiry was consistently shaped by masculine biases. This bias is evident, they argued, not only in the definition of what counts as a scientific problem but also in the interpretations of research. It followed that science could not be genuinely objective until the masculine bias was eliminated. As we shall see below, this approach leaves unchallenged the existing methodological norms of scientific inquiry and identifies only bad science and not science-as-usual as the problem.

The radical political movements of the late 1960s and early 1970s had begun also with the question of the use and abuse of science. In their campaigns against an abused, militarised, and polluting science they argued that science was directed towards profit and warfare. Initially, science itself was seen as neutral/value-free and useful as long as it was in the hands of those working for a just society. Gradually, however, the radical science movement developed a Marxist analysis of the class character of science and its links with capitalist methods of production. A revived political economy of science began to argue that the growth and nature of modern science was related to the needs of capitalist society. Increasingly tied to the state and industry, science had become directed towards domination.

During this same period a radical shift took place in the history, philosophy and sociology of science, which added weight to the view that science could no longer be understood simply as the discovery of reality. Thomas Kuhn's *The Structure of Scientific Revolutions* (1970) marked the beginning of what was to become a major new field of study known as the sociology of scientific knowledge. Its central premise is that scientific knowledge, like all other forms of knowledge, is affected at the most profound level by the society in which it is developed.

Much research has examined the circumstances in which scientists actually produce scientific knowledge and has demonstrated how social interests shape this knowledge. Studies provide many instances of scientific theories drawing models and images from the wider society. It has also been demonstrated that social and political considerations enter into scientists' evaluations of the truth or falsity of different theories. Even what is considered as 'fact', established by experiment and observation, is social. Different groups of scientists in different circumstances have produced radically different 'facts'. Numerous historical and contemporary studies of science, and the social processes through which inquiry proceeds, highlight the social aspects of scientific knowledge.

However, despite the advances that were made through the critique of science in the 1970s, gender-conscious accounts were rare. The social studies of natural science systematically avoided examining the relationship between gender and science in either its historical or sociological dimensions. Similarly, the radical science movement focused almost exclusively on the capitalist nature of science, ignoring the relationship of science to patriarchy. In short, gender did not figure as an analytical tool in either of these accounts of science.

It is only during the last decade with writers such as Ludmilla Jordanova (1980), Carolyn Merchant (1980), Brian Easlea (1981), Elizabeth Fee (1981), Nancy Hartsock (1983), Hilary Rose (1983) and Evelyn Fox Keller (1985), that Western science has been labelled as inherently patriarchal. As Sandra Harding (1986, p. 29) expresses it, feminist criticisms of science had evolved from asking the 'woman question' in science to asking the more radical 'science question' in feminism. Rather than asking how women can be more equitably treated within and by science, they ask how a science apparently so deeply involved in distinctively masculine projects can possibly be used for emancipatory ends. It is therefore time to consider the main feminist critiques of science itself.[2]

Scientific knowledge as patriarchal knowledge

The concern with a gender analysis of scientific knowledge can be traced back to the women's health movement that developed during the 1970s. Regaining control over sexuality and fertility was seen as crucial to women's liberation. Campaigns for improved birth control and abortion rights were central to the early period of second-wave feminism. There was growing disenchantment with male medical theories and practices. Even the imagery employed in medical discourses is strikingly infused with gender stereotypes. Just to give one example, Emily Martin (1991) found that scientific accounts of reproductive biology invariably represent the egg as behaving 'femininely' and the sperm as behaving 'masculinely'. The process of fertilisation is described in terms of the powerful, active sperm penetrating the passively waiting egg. This is not a 'fact' of biology but a particular way of describing the natural world. Indeed recent research suggests that sperm and egg are mutually active partners. Women began to develop new kinds of knowledge and skills which drew on their own experience and needs. The insights of the radical science movement contributed to the view of medical science as a repository of patriarchal values.

If medical scientific knowledge is patriarchal, then what about the rest of science? As Maureen McNeil (1987) points out, it was a short step to the emergence of a new feminist politics about scientific knowledge in general. Many feminists re-examined the Scientific Revolution of the sixteenth and seventeenth centuries, arguing that the science which emerged was fundamentally based on masculine projects of reason and objectivity. They characterised the conceptual dichotomies central to scientific thought, and to Western philosophy in general, as distinctly masculine. Culture versus nature, mind versus body, reason versus emotion, objectivity versus subjectivity. In each dichotomy the former must dominate the latter, and the latter in each case seems to be systematically associated with the feminine.

Rather than pointing to the negative consequences of women's identification with the natural realm, some feminists celebrate the identification of woman and nature. This finds political expression in the radical feminism and eco-feminism of the 1980s which suggests that women must and will liberate the earth because they are more in tune with nature. For them, women's involvement in the ecology and peace movements was evidence of this special bond. Women's biological capacity for motherhood was seen as connected to an innate selflessness born of their responsibility for ensuring the continuity of life.

Conversely, men's inability to reproduce has made them disrespectful of human and natural life, resulting in wars and ecological disasters. From this perspective, a new feminist science would embrace feminine intuition and subjectivity and end the ruthless exploitation of natural resources. Rejecting patriarchal science, this vision celebrates female values as virtues and endorses the close relationship between women's bodies, women's culture and the natural order. A feminist science, in other words, would be based on women's values.

This view of feminist science is not one to which all feminists subscribe, however. Essentialism, or the assertion of fixed, unified and opposed female and male natures has been subjected to a variety of thorough critiques. The first thing that must be said is that the values that are ascribed to women originate in the historical subordination of women. The belief in the unchanging nature of women, and their association with procreation, nurturance, warmth, creativity, lies at the very heart of traditional and oppressive conceptions of womanhood. It is important to see how women came to value nurturance and how nurturance, associated with motherhood, came to be culturally defined as feminine within male-dominated culture. Rather than asserting some inner essence of womanhood as an ahistorical category, we need to recognise the ways in which both 'masculinity' and 'femininity' are socially constructed and are in fact constantly under reconstruction.

The idea of 'nature' is itself also culturally constructed. Conceptions of the 'natural' have changed radically throughout human history. As anthropologists like Marilyn Strathern and others have pointed out, 'no single meaning can in fact be given to nature or culture in Western thought; there is no consistent dichotomy, only a matrix of contrasts' (Strathern 1980, p. 177). These feminist anthropologists have questioned the claim that in all societies masculinity is associated with culture and femininity with nature. Arguments such as these cast serious doubt on the project for a feminist science presented above. Once it is recognised that 'masculinity' and 'femininity', as well as the idea of 'nature', are changing cultural categories then it no longer makes sense to base a science on feminine intuition rooted in nature.

Finally, a general problem with the literature on gender and science is that much of the debate centres on questions of the methodology or epistemology of science. It may be that the search for the most appropriate feminist epistemology, however philosophically sophisticated, is misdirected. The more philosophically oriented feminist work on science suffers from the familiar problem of dealing with ideas divorced from social practices. Indeed, as amply shown by these authors, statements of 'The Scientific Method' do typically contain male visions of what it is to know and what the world is really like. But scientific practice is in no sense determined by statements of method. The latter are better seen as political pronouncements, as legitimations, rather than as descriptions of what scientists actually do.

It is in this light that we should see attempts to spell out a specifically feminist scientific method. They are politically useful in that they turn the feminist spotlight on the content of scientific knowledge instead of simply highlighting questions of recruitment to science. But we need to be cautious in presuming that the adoption of a 'feminist' scientific method would lead to differences in scientific practice without a thoroughgoing change in the relations of power within science. The danger is that what might parade as feminist science would simply amount to the same scientific practice by another name.

In other words, there is a danger that feminists treat descriptions of 'doing science' *literally*. In fact, there is a huge gulf between how science (or engineering for that matter) is said to be done (that is, in a cool, objective, logical, rational way) and how it is really done (with excitement, obsession and passion). The former is a very powerful ideology and serves to position women as unsuitable for these endeavours.

From science to technology

Many feminist approaches to technology mirror those to science outlined above, but they are more recent and much less developed than those which have been articulated in relation to science. As a result, I will draw out strands of argument from this literature rather than present the material as coherent positions in a debate.

An initial difficulty in considering the feminist commentary on technology arises from its failure to distinguish between science and technology. The effect of this is commonly to conflate technology with science. To some extent this is because feminist writing on science has often construed science purely as a form of knowledge, and this assumption has been carried over into much of the feminist writing on technology. However, just as science includes practices and institutions, as well as knowledge, so too does technology. Indeed, it is even more clearly the case with technology, because technology is primarily about the creation of artefacts.

In recent years, there has been a major reorientation of thinking about the form of the relationship between science and technology. The model of the science–technology relationship which enjoyed widespread acceptance over a long period was the traditional hierarchical model which treats technology as applied science. This view—that science discovers and technology applies this knowledge in a routine uncreative way—is now in steep decline. 'One thing which practically any modern study of technological innovation suffices to show is that far from applying, and hence depending upon, the culture of natural science, technologists possess their own distinct cultural resources, which provide the principal basis for their innovative activity' (Barnes & Edge 1982, p. 149). Technologists build on, modify and extend existing technology but they do this by a creative and imaginative process. And part of the received culture technologists inherit in the course of solving their practical problems is non-verbal; it cannot be conveyed adequately by the written word, either. Instead it is the individual practitioner who is the basic unit in the transfer of knowledge and competence.

Hidden from history

Feminists have pointed out that women's contributions have by and large been left out of technological history. The history of technology represents the prototype inventor as male. So, as in the history of science, an initial task of feminists has been to uncover and recover the women hidden from history who have contributed to technological developments. For example, Autumn Stanley's book, *Mothers and Daughters of Invention*, is a history of women inventors. Her work suggests that during the industrial era, women invented or contributed to the invention of such crucial machines as the cotton gin, the sewing machine, the small electric motor, the McCormick reaper, and the Jacquard loom.

To fully comprehend women's contributions to technological development, however, a more radical approach may be necessary. To begin with, the traditional conception of technology too readily defines technology in terms of male activities. The concept of technology is itself subject to historical change and different epochs and cultures had different names for what we now think of as technology. A greater emphasis on women's activities immediately suggests that females, and not males, were the first technologists. It is now well established that they were the main gatherers, processors and storers of plant food from earliest human times. It was therefore logical that they should be the ones to have invented the tools and methods involved in this work.

If it were not for the male bias in most technology research, the significance of these inventions would be acknowledged. As Ruth Schwartz Cowan, an American historian of technology, notes: 'The indices to the standard histories of technology . . . do not contain a single reference, for example, to such a significant cultural artifact as the baby-bottle. Here is a simple implement . . . which has transformed a fundamental human experience for vast numbers of infants and mothers, and been one of the more controversial exports of Western technology to underdeveloped countries—yet it finds no place in our histories of technology' (1979 p. 52).

There is, therefore, important work to be done not only in identifying women inventors, but also in discovering the origins and paths of development of 'women's sphere' technologies that seem often to have been considered beneath notice. In other words, we need to try to undo the links in this association between what technology is and what men do. The cultural stereotype of technology as an activity appropriate for men is an issue I will return to. Before doing that, however, I want to consider another argument that is increasingly popular.

Technology based on women's values?

During the 1980s, feminists have begun to focus on the gendered character of technology itself. Rather than asking how women could be more equitably treated within and by a neutral technology, many feminists now argue that Western technology itself embodies patriarchal values. Technology, like science, is seen as deeply implicated in the masculine project of the domination and control of women and nature. And just as many feminists have argued for a science based on women's values, so too has there been a call for a technology based on women's values. In Joan Rothschild's preface to a collection on feminist perspectives on technology, she says that: 'Feminist analysis has sought to show how the subjective, intuitive, and irrational can and do play a key role in our science and technology' (1983 p. xxii).

Male authors have also advocated a technology based on women's values. Mike Cooley in *Architect or Bee?* (1980, p. 43) argues that technological change has male values built into it: 'the values of the White Male Warrior, admired for his strength and speed in eliminating the weak, conquering competitors and ruling over vast armies of men who obey his every instruction . . . Technological change is starved of the so-called female values such as intuition, subjectivity, tenacity and compassion'.

Cooley sees it as imperative that more women become involved in science and technology to challenge and counteract the built-in male values: that we cease placing the objective above the subjective, the rational above the tacit, and the digital above analogical representation. Similarly, in *The Culture of Technology,* Arnold Pacey (1983) devotes an entire chapter to 'Women and Wider Values'. He outlines three contrasting sets of values involved in the practice of technology—first, those stressing virtuosity; second, economic values; and third, user- or need-oriented values. Women exemplify this third 'responsible' orientation, according to Pacey, as they work with nature in contrast to the male interest in the conquest of nature!

Ironically the approach of these male authors is in some respects rather similar to the eco-feminism that became popular among feminists in the 1980s. This marriage of ecology and feminism rests on the 'female principle', the notion that women are closer to nature than men and that the technologies men have created are based on the domination of nature in the same way that they seek to dominate women. Eco-feminists concentrated on military technology and the ecological effects of other modern technologies. According to them, these technologies are products of a patriarchal culture that 'speaks violence at every level'. However, an inevitable corollary of this stance seemed to be the representation of women as inherently nurturing and pacifist. The problems with this position have been outlined above in relation to a science based on women's essential values. We need to ask how women became associated with these values.

This strand of argument has been very positive, however, in taking the debate about gender and technology beyond the use/abuse model of technology and focusing on the political qualities of technology itself. It has also been a forceful assertion of women's interests and needs as being different from men's and has highlighted the way in which women are not well served by current technologies. And, finally, it has contributed to a much more sophisticated debate about women's exclusion from the processes of innovation and from the acquisition of technical skills.

Feminists have pointed to all sorts of barriers—in social attitudes, girl's education and the employment policies of firms—to account for the imbalance in the number of women, for example, in engineering. But we must also ask whether women actively resist entering work in technology. Why have the women's training initiatives designed to break men's monopoly of the building trades, engineering and information technology not been more successful? Although schemes to channel women into technical trades have been small-scale, it is hard to escape the conclusion that women's response has been tentative and perhaps ambivalent.

Technology and the division of labour

These issues can best be dealt with by an historical and sociological approach to the analysis of gender and technology. This approach has built on some theoretical foundations provided by contributors to what is known as the labour process debate of the 1970s. The basic argument of the labour process literature was that capitalist–worker relations are a major factor affecting the technology of production within capitalism. Historical case studies of the evolution and introduction of particular technologies documented the way in which they were deliberately designed to deskill and eliminate human labour. So, like science, technology was understood to be the result of capitalist social relations.

This analysis paved the way for the development of a more sophisticated analysis of gender relations and technology. However, the labour process approach was gender-blind because it interpreted the social relations of technology in exclusively class terms, ignoring the argument that the relations of production are constructed as much out of gender divisions as class divisions. Recent feminist writings in this historical vein see women's exclusion from technology as a consequence of the male domination of the skilled trades that developed during the industrial revolution.

Male craft workers could not prevent employers from drawing women into the new spheres of production. So instead they organised to retain certain rights over technology by actively resisting the entry of women to their trades. Women who became industrial labourers found themselves working in what were considered to be unskilled jobs for the lowest pay. Thus male dominance of technology has largely been secured by the active exclusion of women from areas of technological work. This gender division of labour within the factory meant that the machinery was designed by men with men in mind, either by the capitalist inventor or by skilled craftsmen. Industrial technology from its origins thus reflects male power, as well as capitalist domination.

The masculine culture of technology is fundamental to the way in which the gender division of labour is still being reproduced today. By securing control of key technologies, men are denying women the practical experience upon which inventiveness depends. Segregated at work and primarily confined to the private sphere of the household, women's experience has been severely restricted and therefore so too has their inventiveness. An interesting illustration of this point lies in the fact that women who were employed in the munitions factories during World War I are on record as having redesigned the weaponry they were making. Thus, given the opportunity, women have demonstrated their inventive capacity in what now seems the most unlikely of contexts.

Missing: the gender dimension in the sociology of technology

The historical approach is an advance over essentialist positions which seek to base a new technology on women's innate values. Women's profound alienation from technology is accounted for in terms of the historical and cultural construction of technology as masculine. Women's exclusion from and rejection of technology is made more explicable by an analysis of technology as a culture that expresses and consolidates relations among men. If technical competence is an integral part of masculine gender identity, why should women be expected to aspire to it?

Such an account of technology and gender relations, however, is still at a general level. There are few cases where feminists have really got inside the 'black box' of technology to do detailed empirical research, as some of the most recent sociological literature has attempted. Over the last few years, a new sociology of technology has emerged which is studying the invention, development, stabilisation and diffusion of specific artifacts. The work represented in, for example, *The Social Construction of Technological Systems* (edited by Bijker et al. 1987), known as the social shaping or social constructionist approach comes to mind. It is evident from this research that technology is not simply the product of rational technical imperatives. It attempts to show the effects of social relations on technology that range from fostering or inhibiting particular technologies, through to influencing the choice between competing paths of technical development, to affecting the precise design characteristics of particular artifacts.

Because social groups have different interests and resources, the development process brings out conflicts between different views of the technical requirements of the device. Accordingly, the stability and form of artifacts depends on the capacity and resources that the salient social groups can mobilise in the course of the development process. Thus, in the technology of production, economic and social class interests often lie behind the development and adoption of devices. In the case of military technology, the operation of bureaucratic and organisational interests of state decision making will be identifiable. Indeed, state sponsorship of military technology shapes civilian technology.[3]

So far, however, little attention has been paid to the way in which technological objects may be shaped by the operation of gender interests. This blindness to gender issues is also indicative of a general problem with the methodology adopted by the new sociology of technology. These writers use case studies of innovation as models to show how social groups actively seek to influence the form and direction of technological design. What they overlook is the fact that the absence of influence from certain groups may also be significant. For them, women's absence from observable conflict does not indicate that gender interests are being mobilised. For a social theory of gender, however, the almost complete exclusion of women from the technological community should alert us to the underlying structure of gender relations. Preferences for different technologies are shaped by a set of social arrangements that reflect men's power in the wider society. The process of technological development is socially structured and culturally patterned by various social interests that lie outside the immediate context of technological innovation.

The gendered relations of workplace technology

One of the most important ways that gender divisions interact with technological change is through the price of labour, in that women's wage labour generally costs considerably less than men's. This may affect technological change in at least two ways. Because a new machine has to pay for itself in labour costs saved, technological change may be slower in industries where there is an abundant supply of cheap women's labour. The classic example of this is the clothing industry which has remained technologically static since the nineteenth century. While there are no doubt purely technical obstacles to the mechanisation of clothing production, there will be less incentive to invest in automation if skilled and cheap labour is available to do the job. Thus there is an important link between women's status as unskilled and poorly paid workers, and the uneven pace of technological development.

There is a more direct sense, however, in which gender inequality leaves its imprint on technology. Employers may seek forms of technological change that enable them to replace expensive skilled male workers with poorly paid female workers, less likely to be unionised. A good example of this comes from Cynthia Cockburn's (1983b) account of an archetypal group of skilled workers being radically undermined by technological innovation. It is the story of the rise and fall of London's Fleet Street compositors, an exclusively male trade with strong craft traditions of control over the labour process. A detailed description of the technological evolution from the Linotype system to electronic photocomposition shows how the design of the new typesetting technology reflected gender relations.

The new computerised system was designed with the keyboard of a conventional typewriter rather than the compositor's traditional, and very different, keyboard. There was nothing inevitable about this. Electronic circuitry is in fact perfectly capable of producing a Linotype keyboard lay-out on the new-style board. So what politics lie behind the design and selection of this keyboard? In choosing to dispense with the Linotype layout, management were choosing a system that would undermine the skill and power basis of the compositors, and reduce them at a stroke to 'mere' typists. This would render typists (mainly women) and compositors (men) equal competitors for the new machines: indeed, it would advantage the women typists. The new typesetting technology was designed with an eye to using the relatively cheap and abundant labour of female typists.

This demonstrates very well that although machine design is overwhelmingly a male province, it does not always coincide with the interests of men as a sex. Gender divisions are commonly exploited in the power struggles between capital and labour. In this case, and it is not unique, a technology was designed for use by women in order to break the craft control of men. Cockburn's study is a good illustration of the extent to which gender relations as well as capitalist relations are built into industrial technology.

Technologies reflect the distribution of power and resources in society and, in a society such as ours which is characterised by gender inequality, they reflect men's interests much more than women's. This is a complex argument, however, because there is no one simple theory of the way technologies are shaped by gender relations. Rather, in looking at different types of technology, it can be seen that men's interests shape them in different ways and to varying degrees. To understand any specific process or product, we must ask: who developed it, and why; in whose interest? To illustrate this further, contrast the above account of the shaping of a workplace technology with the development of domestic technologies.

The gendered relations of domestic technology

We usually think of the home as a technology-free zone but in reality it is a hive of technological activity. Most households now sport a bewildering array of appliances, from washing machines and dishwashers to food processors, refrigerators, vacuum cleaners and microwaves. Many of these are sold to us as labour-saving—but in fact women spend nearly as much time on housework as our grandmothers did. This is partly because standards of housework have risen and partly because the technologies sold for the home are not so efficient. Why is it that cookers and vacuum cleaners are so inefficient and yet moon landings and so-called 'smart' weapons are possible? Domestic technologies are clearly intended to be used by women, yet women's interests do not inform the design process.

The fact is that much domestic technology was not specifically designed for household use but has its origins in very different spheres. Many appliances were initially developed for commercial, industrial or even military purposes and only later were they adapted for home use. Microwave ovens, for example, are a direct descendant of military radar technology and were developed for food preparation in submarines by the US Navy. They were first introduced to airlines, institutions and commercial premises before manufacturers turned their eyes to the domestic market. For this reason new domestic appliances are not always appropriate to the household work that they are supposed to perform. Nor are they necessarily the implements that would have been developed if the housewife had been considered first or, indeed, if she had control of the processes of innovation.

One such example is the self-cleaning house—every woman's impossible dream (Zimmerman, 1983). Frances Gabe, an artist and inventor from Oregon, spent twenty-seven years building and perfecting the self-cleaning house. In effect, a warm water mist does the basic cleaning and the floors (with rugs removed) serve as the drains. Every detail has been considered. 'Clothes freshener cupboards' and 'dish washer cupboards' which wash and dry, relieve the tedium of stacking, hanging, folding and ironing and putting away. Gabe was ridiculed for even attempting the impossible, but architects and builders now admit that her house is functional and attractive. One cannot help speculating that the development of an effective self-cleaning house has not been high on the agenda of male engineers.

There are economic considerations, of course. Women's domestic labour is unpaid. As an industrial designer I interviewed said: 'Why invest heavily in the design of domestic technology when there is no measure of productivity for housework as there is for industrial work?' Instead, when producing for the household market, manufacturers concentrate on cutting the costs of manufacturing so that they can sell reasonably cheap products. Much of the design effort is put into making appliances look attractive or impressively 'high-tech' in the showroom—giving them for example an unnecessary array of buttons and flashing lights. Far from being designed to accomplish a specific task, some appliances are designed expressly for sale as moderately priced gifts from husband to wife and, in fact, are rarely used. In these ways the inequalities between men and women, and the subordination of the private to the public sphere, are reflected in the very design processes of domestic technology.

Conclusion

I have argued that a gendered approach to technology cannot be reduced to a view which treats technology as a set of neutral artifacts manipulated by men in their own interests. Rather, the social shaping approach insists that technology is always a form of social knowledge, practices and products. It is the result of specific decisions made by particular groups of people in particular places at particular times for their own purposes. The focus here is on how gender interests shape technology, but of course there are other powerful forces such as militarism and capitalist profitability.

Developing a theory of the gendered character of technology is fraught with difficulty. On one hand, there is a risk of adopting an essentialist position that sees technology as inherently patriarchal. On the other, it is easy to lose sight of the structure of gender relations through an over-emphasis on the historical variability of the categories of 'women' and 'technology'. This now familiar feminist tightrope is best regarded as a creative tension. In any event, I would argue that an attempt to construct a general feminist theory of technology is premature at this stage. What we need are more case studies analysing the specific social interests that structure particular kinds of technological innovation.

The emphasis in empirical work in the social shaping tradition to date has been on manufacturing technologies, and gender issues have been marginalised by the way mainstream research has been conceived and executed. This is partly because a gender perspective is seen as being of relevance only when the research specifically addresses women. This allows people to think that they do not need to deal with gender issues if they are not researching women. Bringing a gender awareness to research involves seeing gender relations as a significant explanatory concept—and treating both men and women as having a gender identity which structures experiences and understandings. As I indicated at the outset, feminist research on workplace technologies has focused on the office. It has been mainly concerned with the effects of technological change on a predominantly female workforce. This book includes excellent examples of research which not only examine the impact of new technologies but also recognise that gender relations are embodied in the design of office technologies. Research on office technologies adds a crucial dimension to our understanding of the relationship between work, gender and technology.

Notes

[1] This article draws on Chapter 1 of my book *Feminism Confronts Technology*, Sydney, Allen and Unwin; Cambridge, Polity Press; Pennsylvania, Penn State University Press, 1991.

[2] It is important to understand that, by presenting these developments in a chronological order, I do not mean to imply that earlier projects are not ongoing and extremely valuable. The efforts of those—including myself—involved in equal employment opportunity work, increasing women's participation in computing and other scientific and technical occupations, are not misconceived, although they may have limited potential.

[3] Much of the research on electronics in this country has been sponsored by the military, especially in the United States. Military exigencies and military support have been crucial in the development of 'civilian' technologies, such as the digital computer. See David Dickson, *The New Politics of Science*, Pantheon, New York, 1984, for a description of the ever-closer relationship between science institutions and military–industrial interests in the USA.

3

Women's skills and word processors

Gender issues in the development of the automated office

JULIET WEBSTER

This chapter looks at the social relations within offices and how they have provided the basis for the development of office automation.[1] It starts from the premise that the evolution and form of technologies are governed not only by advances in techniques and equipment, but also by a whole host of economic, political, social, cultural and organisational factors.

The historical background to what is known as the 'social shaping' approach to the study of technology has been discussed in the preceding chapter by Judy Wajcman. We now know well that technological determinism—the idea that technological development is simply a series of 'objective' technical improvements—has much to be criticised. So we have had studies of the social shaping of machine tools, of robots, of electricity systems, of telecommunications switching systems, of software, of bicycles, and of nuclear missile guidance systems. For example, Hughes's (1983) research into Edison's invention of the electric light bulb has shown that technological artifacts are typically as much the outcome of economic decisions as they are of technological decisions. Similarly, Noble's well-known (1979) study of the evolution of machine tools highlights the control and deskilling of engineering labour as a key motivating force in the development of numerical control in preference to record playback machine tool technology. For him, the ideology and practice of shopfloor control directly reflects the antagonistic social relations of capitalist production;

thus, the design of machinery, informed by this ideology, reflects the social relations of production.

As Wajcman points out, however, few studies have been informed by a consideration of how gender relations condition the development of technologies. Relations of power and inequality between men and women, but also between women and women, and between men and men inform the structure and culture of social institutions and the interaction between their members. Gender relations constitute a key, and often neglected, part of the social arrangements within which technologies evolve. It is this analysis of how gender relations affect the development of a workplace technology which I attempt in this chapter. Specifically, I show how the development of a particular component of office automation—the word processor—can be seen as crucially linked to the gendering of office work and of the typing task, and in particular of the value that is attributed to the work that women office workers do.

However, I am not simply concerned to make 'gender relations' one more weapon in the armoury of social factors shaping the design of technologies, and thereby undermining the concept of technological determinism. I want to show how office technologies (with this gendered design) are introduced into workplace settings where gender relations, gendered divisions of labour, and gendered expertise also operate, which are affected by, as well as affecting, these technologies. In other words, I emphasise the mutuality of the relationship between technical and organisational change, and show how changes in technology are socially shaped, socially shaping, and, as a result of their shifting social contexts, continually evolving and being redesigned.

The chapter therefore starts by considering the development of word processing technology in terms of the social structures, and in particular the gender relations, within which it has emerged. It then examines the impact of word processing systems upon the area of office work most fundamentally and visibly affected, namely, women's secretarial and typing work. What were the expectations and, more importantly, what was the actual experience of word processing? How were gender relations in the office affected? Finally, what are the implications for women office workers and what can be done to improve their working conditions?

The evolution of the word processor

Word processing is the outcome of a number of different innovations taking place in different fields of electronic text production and reproduction. Its origins are threefold and distinctive: data processing, programming, and office equipment.

In the early business systems of the 1950s and 1960s, computing was carried out on large-scale mainframe machines which were used entirely for data processing applications such as company accounts or wages and salaries. There were both software and hardware constraints to extending the domain of these computing giants to cover other business applications involving the processing of text. Firstly, screen editing was not possible at that time; all code and data was line-edited. Line editing involved writing a small programme which specified the line and, within that, the character or characters that needed to be altered, and what they should be altered to. This was a tedious process, which militated against the wholesale inclusion of word processing as an application on mainframe machines. Moreover, word processing is highly resource-hungry, involving heavy usage of disk and memory space to do the writing and rewriting, saving and re-saving which is typically involved. Word processing would thus have to compete with the more established data processing applications for meagre computing resources. This was the technical constraint against word processing on mainframes.

Besides this technical rationale there also seems to have been an exercise in social choice for data processing being favoured over text processing applications on mainframes. There was certainly an ethos that computers were a 'hard' mathematical technology, designed to perform strictly numerical tasks. After all, they were originally built for calculating and code-cracking (the Babbage difference engine and the Turing machine, for example), rather than for composing. In this sense, their development, like that of machine tools, can be seen as a reflection of the culture of engineering which identifies itself with the quantitative, rational elements of work. Most early industrial applications of computers involved their calculative, algorithmic and file-handling functions with financial and accounting applications. Initially, these took precedence over activities which were less quantitative in nature, such as the processing of text. Perhaps text processing was seen as a 'soft' application, not a proper utilisation of computing resources, in that it was non-numerical.

In addition to these cultural factors shaping the domain of computing, there is an economic consideration. Word processing—the entering of text—was defined as 'women's work', and as such was performed using relatively cheap labour, never a priority for managements' automation programmes. When word processing did finally develop as a computing application, it was not in the form of full-screen editing and manipulation of text files that we know today. Rather, text processing packages were merged with files of raw text, in order to format them in letter or document quality. This solution allowed word processing to be performed in a way that did not consume inordinate amounts of disk and memory space, and it did not divert resources away from the original, mathematical applications of computing.

Word processing also evolved in the computer programming sphere. The early programmers who were writing software in machine code—a low-level computer language consisting of zeros and ones—needed some kind of tool for editing this code easily and without error. A package was developed for this purpose, which became what we now know as Word-Star. It was not originally meant for office use at all, or for word processing proper. It was only with the diffusion of microcomputers in the 1970s that Word-Star become marketed and used predominantly as a full word processing system with wider applications than simply editing code.

The third line of descent of word processing is not computing at all, but office equipment. From the mid-1950s onwards, office technologies began to undergo a series of enhancements to incorporate increasingly sophisticated facilities for the production of documents, a process which was led by the then IBM Electric Typewriter Division. In 1964 the Magnetic Tape/Selectric Typewriter (MT/ST) was produced: this device was essentially a typewriter which stored text on tape. With the conventional manual typewriters the typist's keystrokes committed the text directly on to paper. The MT/ST stored text on tape before committing it to paper. The slight delay between keying and printing allowed the typist to create rough drafts very rapidly, backspacing and correcting errors as they were made. IBM labelled the concept behind this technology 'power typing', and from the mid-1960s onwards it introduced a range of products in its magnetic media line. Some of these incorporated basic programmes for formatting text, as well as storage of the text itself. Other facilities included justification (positioning of text flush to left-hand or right-hand margins) and centering of text. Other manufacturers, meanwhile, experimented with cathode ray tube display workstations, and in 1972, the development of daisywheel printer technology provided letter quality printing at high speeds and thus increased the popularity of these putative word processing devices. In 1974, the floppy disk, already in general use for computing applications, was imported to word processing as the storage medium. These combined innovations brought into being the dedicated word processor technology which became widely diffused in the offices of the 1970s and early 1980s.

The social shaping of word processing

Why was it that the word processing technology which emerged from office equipment development dominated first generation word processing? Why did this particular technological form become prevalent over others? The question cannot be understood simply in terms of the 'logic' of straightforward technical advance or efficiency, a logic which tends to be taken for granted in technologically determinist accounts. This does not explain why, initially, word processing took the form that it did. This latter form of word processing—with its roots in office equipment rather than in computing—in fact used one of the lowest density, slowest and therefore least efficient character recording schemes available at the time; even by the standards of the 1960s, this specification was extremely 'low tech'. The real significance of this development was that it addressed the productivity of the typist in a form that resembled but enhanced the manual typewriter, and was therefore reasonably familiar to the user. This suggests that in order to understand the logic of the development of word processing, we have to look beyond the technical inputs into these devices, to the social context of their development, to the work processes which they were designed to address, to the characteristics of their users.

The development and diffusion of a particular word processing design that was not the most technically efficient or advanced solution has to be seen in relation to the work processes involved in text production and of the gender of those who do this work. The first typewriters were introduced into offices in the 1870s, and the entry of women, facilitated by the new category of work created, followed shortly after. Since that time, text production has traditionally been performed by women. The most marketable and easily accepted word processor design was likely to be one that was familiar to, and usable by, women office workers. Early word processors therefore followed the design of the typewriter, and extended typewriter technologies. They were specialised machines designed for use by trained touch-typists, with QWERTY keyboards (so called because the keys are laid out with the letters Q, W, E, R, T and Y on the left-hand side of the top row) and an extra pad of specialised command keys labelled with their functions ('Insert', 'Delete' and 'Save' keys combined with 'Word', 'Line', 'Paragraph' and 'Document' keys which could, for example, save a document at the press of two dedicated keys). Many word processors also incorporated a screen design which was intended to represent the sheet of paper. The sales pitch of the equipment manufacturers showed glamorous secretaries in air-conditioned offices, surrounded by pot plants and smiling into visual display units. This form of word processing, then, emulated the typewriter, in order to make the new technology as acceptable and appealing to the typist—and her boss—as possible. In its operation and design, it was a relatively minor departure from the old technology, for it incorporated office activities done traditionally by women into its construction and was designed for use specifically by women office workers (though it was no less daunting to its first users, for all that).

The gendering of typing work has therefore been an important factor in the development of word processing. The form of word processing has been strongly shaped by those who perform this work. This association of machines with operators is not new. When typewriting technology emerged at the end of the nineteenth century, the women who operated them bore the same name as the machines—'typewriters'. The technology rapidly became synonymous with those who performed the work, and, as we have seen, the developing strong association of women with this function ensured that typewriting became a highly gender-specific activity. It was in the context of this gender-specific task that word processing technology emerged, and the term 'word processor' has also sometimes been used to refer to the incumbents of such office jobs as well as to the devices upon which they work. Because this kind of office labour was historically almost exclusively female, this gendered division of labour became incorporated in the technology just as much as the more tangible elements of hardware were part of its design. It meant that a particular hardware solution gained strong currency despite its technical inefficiency. This type of word processor was designed and marketed for use by secretaries and typists trained to touch-type on the QWERTY keyboard and to lay out documents neatly on paper. In this way the division of labour in the office, a sexual division by now built upon 'nimble fingers' and manual dexterity which were the almost exclusive province of women, was perpetuated.

The spread of word processing and the 'office of the future'

How, then, did word processing, shaped as it has been by the sexual division of labour and by the association of women with text production, in its turn affect that division of labour? Has it prompted any change in the composition and performance of office tasks, or in those who perform them? Has text production been deskilled by the move to word processing? How, if it all, has word processing altered the culture of the office and the relationships between office workers? And has the value ascribed to the work that women do in the office been raised in the light of its apparently greater technical content, or lowered in the light of its deskilling?

When word processing began to diffuse into offices in the late 1970s, gloomy predictions were made about its likely impact on women, their work and their skills. It was commonly argued that this technology would bring the regimentation of the assembly line into the office, divesting women of all the skills and abilities which they used in the course of their work (see, for example Barker & Downing 1980). Word processing, it was argued, was a mechanism for the introduction of Taylorism into the office and as such, would be associated with the degradation, deskilling and intensification of office work. Women's work would be even more devalued and women would become mere pressers of buttons.

Since these forecasts were made, however, the experience of information technology in the office has proved quite different. First, it has proved impossible to talk of the 'impact' of word processing in isolation from the office contexts into which it is introduced. Indeed, to talk of 'impacts' would be to fall back into the trap of technological determinism, in which technologies are treated as independent from the social arrangements within which they are located. As we have seen, word processing is not a purely technical artefact, but one which already embodies the sexual division of white-collar labour.

Second, research which has examined the content of women's office labour and skills associated with the move to word processing has shown these to be highly variable, and closely related to the characteristics of work organisation prevailing in particular firms *before* the introduction of word processing (Silverstone & Towler 1983; Webster 1990a). That is to say, the patterns of work organisation and social relations in place in offices before the introduction of word processing have been much more decisive in shaping the character of office jobs after the implementation of this new technology than has the automation process itself. And since these patterns and relations have varied from one office to another, so too has the resultant character of the automated office.

Instead of all office workers being reduced to uniform deskilled automatons, the differential between secretarial and typing workers, and the ways in which their respective jobs are organised, persists. As Arnold et al. rightly note:

> The distinction between secretarial and typing work is crucial to analysis of the likely effects of WP in offices . . . Female typists—essentially machine operators—have existed as long as the machines themselves. As specialist machine operators, they can be managed like factory shopfloor workers; this may not be true for secretaries, who perform a range of other duties in addition to typing (1982, p. 60).

Indeed, secretarial workers continue to do a variety of tasks, with considerable discretion and technical skill, working very much at their own pace and according to their own pressures. Typists, whose work has always been relatively routinised, continue to perform a much smaller range of more fragmented tasks than secretaries have, with much less discretion and much more external direction (often from a typing pool supervisor). Certain kinds of employers, financial institutions in particular, have very long-standing systems of work rationalisation which predate, and actually form the basis for, the introduction of office automation systems. For example, the building societies of the 1970s already had highly automated, highly structured labour processes. They regarded themselves as having an organisational system of word processing in place. Correspondence was routinised and simplified: pads of pre-printed letters with tick boxes were used, and there were strict divisions between clerical staff and the typing pool for handling different categories of work. One building society Organisation & Methods Manager, responsible for the design of work and systems within the company, simply saw the company's word processing system as 'an accumulation of what had gone on before'. By this he meant that the organisational rationalisation of work had taken place over a long period and considerably before the change in technology which we now recognise as word processing.

Many office workers of all categories continue to bring considerable expertise and competence—of both a technical and broader organisational nature—to bear in the course of their work, and these have by no means been eradicated by the introduction of new technologies into the office. It is the case, though, that these competences often go unrecognised by the observers of office labour processes, including members of management who are unfamiliar with the detail of these labour processes. Women workers themselves also tend to understate their skills and the complexity of the work they do, an issue which is addressed by Cate Poynton in her contribution to this book. It is therefore very easy for women's abilities to be downgraded and denigrated, not just by their superiors, but also by academic commentators anxious to demonstrate the degradation to which *all* automated work is subject.

The competences that women office workers bring to bear in their work vary from the organisational skills of the secretary who ensures the smooth running of the office and deals with the office politics which are never explicit, to the typist or word processor operator who, despite the most routine of jobs, still knows how to get the best from her machine and the workings of her particular office environment. The secretary who effectively runs the office without, predictably, any recognition, is a familiar phenomenon. Certainly, this may consist of doing a great number of domestic tasks which typify the work of 'the office wife'. However, through the administrative side of their work—dealing with phone callers, arranging internal meetings, processing salaries and pensions, and acting as gatekeeper for their bosses—they often develop a strong familiarity with, and sometimes influence over, the running of the firm. Women usually have this influence in the 'private domain' (Stacey 1960), but secretaries can be influential in the 'private-within-the-public' domain of the office. The secretaries in my study of word processing in Britain were constantly going round the office to arrange meetings and appointments with other managers; one preferred to do this personally because of the smallness of the company, which meant that she quickly became drawn into office politics and negotiation processes between different members of management (Webster 1990). She became a repository of organisational knowledge, and soon people stopped in her office, en route to visit her boss, whose office was next door to hers, when they wanted to discuss certain aspects of company business or to exchange information. This organisational expertise and influence has been well described by Vinnicombe in her study of secretarial work:

> At the top of organisations, where secretaries traditionally operate in the one-to-one working relationship with their managers, secretaries carry out a variety of administrative tasks. These tasks all tend to stem from the secretary's gatekeeper position in her boss's communication network. Theoretically, this position gives the secretary almost complete control over the boss's communications. It also means that she has the opportunity to wield a great deal of influence. The extent to which she can influence matters . . . is also related to her personality and the number of years she has worked for the organisation. The last point is important and frequently underestimated. Many top secretaries have well-developed personal contacts throughout the organisation and have an extensive knowledge of the organisation's activities—and sometimes even its secrets (1980, p. 105).

Even the most routine of office workers, without the level of autonomy and discretion that characterises secretarial or supervisory work, commonly know far more than their superiors about the organisation of the office, the labour process and, within this, the operation of technologies. Superiors, without understanding what this work entails, commonly undervalue it. One woman, who worked in a word processing pool overseen by a male boss who thought that by seeing he would be able to understand everything that went on in the office, commented:

> Terry thinks he knows, but he hasn't a clue what goes on. He isn't going to hear this, is he? They think it is easy [working the word processor]. They think you just press a button and that is it. It takes a lot more doing than they think. It still has to be typed out the same. They just think you press a button and it types it itself. Terry even thinks it types on its own.

Fed up with being constantly undervalued and oversupervised, a group of typists decided to embark on a campaign to illustrate how difficult their work was made by poor quality audio dictation, and thus to demonstrate what unrecognised competences they actually required:

> If they said 'er', we put 'er' in. Or they had a bad habit of saying, 'Dear Sir, We have none of your pistons in stock, number RA247 . . .'. And then they say, 'Oh, no. We have *some* of your pistons . . .'. That is no good because you have already typed it if you are on a typewriter. You are only about one word behind all the time. So what we did was, we typed exactly what they said. You just can't correct it when you are on a typewriter; you have got to pull the whole thing out and start afresh when they go and say 'Oh no' and then dictate another sentence.
>
> So some of them came back and were furious about this and said 'This is absolute rubbish'. We said, 'Well, that is what you dictated, so we typed it'. They wouldn't admit that they had done it . . . And they do. They are really bad. They say 'Oh no, typist', so we typed 'Oh no, typist'. They didn't like that at all.

Both managements and radical critics of the 'office wifely' aspects of secretarial work have contributed to the denigration of the skill or knowledge involved in some office jobs. Managements vest intelligence in themselves or in machines, usually because, as the anecdote above indicates, they have little understanding of the actual processes involving in performing certain office tasks, or of the competences required. Radical critics, too, over-influenced by both managerial ideology and Bravermanian arguments about the lack of skill required to perform tasks in the modern office, attribute little if any skilled activity to the office worker. While it is important not to romanticise the abilities required to operate the word processing technologies of today's offices or to perform general secretarial tasks, it is also the case that women office workers themselves clearly vest in their activities *some* levels of skill and expertise. Secretaries and typists commonly express frustration with attitudes which rob them of any recognition of their capabilities.

Part of the problem lies in the fact that both managements and labour process writers have tended to define 'skill' in very conventional terms, in relation to the characteristics of *male* craft labour. Women's labour, no matter how much technical dexterity, mental expertise or training it requires, is usually defined as inferior simply because it is women's labour (Phillips & Taylor 1980; Wajcman 1991). Women's skills have come to be defined as non-technical and so undervalued because, unlike skilled male workers, women have historically held little bargaining ability with which to secure recognition of the competencies they possess. Definitions of 'skill' and technical know-how have become inextricably tied up with the bargaining power of particular male workers, and have also provided a means of excluding women from male preserves of that power. Moreover, ideas about what is 'skilled' work are not only shaped by the bargaining muscle of those who perform it; know-how and technical competence are resources that in their turn confer potential or actual power. This creation and maintenance of power by men to the exclusion of women has been central to the sexual and class politics of technological work (Cockburn 1985). Yet, as Rosemary Pringle (1988) points out, the skills and interpersonal relations and even power involved in women's office work, particularly secretarial work, make it difficult to define their status and position in the office. Certainly they cannot be regarded as simply unskilled or lacking in expertise.

The unrecognised knowledge wielded by women office workers may give them unexpected, but not always positive, elements of control over the functioning of the office. A typing pool supervisor in my British study spent most of her time trying to foster the illusion that the pool was efficiently churning out its work. Unfortunately, the typists were so poorly trained and so ignorant of the overall workings of the organisation that their standard of work was very low, and as a result, the users of the pool were losing confidence in it and doing as much of their own typing as they could. The supervisor's response to this vicious cycle was to give the typists only the most simple and undemanding tasks, and to send the important or complex work out to a typing bureau. At the same time, the typists had to be seen to be looking busy, so she would make them duplicate each other's batches of letters (without telling them that this was happening). Moreover, the word processor which the company had bought stood idle in the corner of the typing pool; its use could not be readily incorporated into this subversive system of work organisation.

Despite the assumptions frequently made by their male superiors that women office workers have only trivial expertise, and the assumptions of many academics that information technology would divest them of any expertise, secretaries and typists do continue to exercise both organisational and technical competence, as we have seen. In fact, particularly in relation to office technologies, women possess much greater competence than their male colleagues and superiors—something which is not surprising given their greater familiarity with the day-to-day operation of these devices.

In the days of the photocopier, the filing cabinet, the typewriter and even the dedicated word processor (highly gendered technologies from which men kept their distance), this was not a problem. Men had no desire to become familiar with these implements, for there were specialised, subordinate office workers for this purpose. Indeed, men fostered their own ignorance of these technologies in order to successfully maintain this distance, eschewing, for example, the operation of keyboards lest they be seen to performing a 'low-grade' function. An item on the BBC Radio 4 'Today' programme on Monday 5 March 1990 illustrated the ignorance of a man and the competence of his secretary, in relation to office technologies and overall procedures. The item concerned Jim Hodgkinson, a manager of a branch of B&Q, the do-it-yourself chain, and Julie Andrews, his PA, who swapped jobs (though not salaries or promotion prospects) for a day. Whereas Julie found no difficulty in taking over Jim's job, simply executing the tasks which previously she left already prepared for him to execute, he was totally unable to cope with hers. It was the technologies that defeated him particularly. 'The technology beat me, I'm afraid', he said. 'The word processors and facsimile machines and technology dotted around the place were just mind-boggling.' With the increasing diffusion of computerised information technologies into workplaces of all kinds, this sort of ignorance has become an increasing problem for male office workers.

Word processing in the 1990s: shifts in the gendered division of labour in the office

With the spread of microcomputers and business packages like spreadsheet systems Visicalc and Lotus 123, the all-purpose office computer has become increasingly popular. There has been a long-established concentration of power and expertise in the hands of remote professionals within the computing specialism. However, over the last decade, managerial end users have challenged the control of centralised data processing departments in organisations, demanding greater independence from erstwhile information technology professionals and pressing for greater autonomy over their own local and individual information handling activities. The growing popularity of microcomputers and their applications can be seen partly as a function of this struggle over the domain of computing activities and expertise. An increasing range of business applications has been made available for 'stand-alone' microcomputers, including word processing packages like Word-Star (which, as we have seen, initially emerged in a quite different context and for a quite different application but have markedly improved in sophistication). As this convergence of information technology applications into a single device has occurred, so the popularity of the specialised word processing machine which is dedicated to one function only, has declined. Since the mid-1980s, dedicated word processing machines as separate pieces of hardware have been growing increasingly rare and are forecast to disappear altogether. The trajectory of the development of word processing has changed.

With the rise of the all-purpose office computer, the specific and exclusive association of keyboard operating with degraded women's work has become weaker. As word processing has become associated less with dedicated electronic replicas of typewriters and more with general computing, it has become more acceptable for men to use these devices with dexterity. No longer is keyboard operating the sole province of women office workers. On the contrary, courses in keyboard skills are now specifically designed to attract male managers. For example, Sight And Sound, an organisation which runs typing courses, has launched 'Breakfast Time Training for the Busy Executive', a four-week course in keyboard skills aimed at 'putting the businessman in control of his (sic) computer' (as reported in the *Scotsman*, 20 February 1990). Even secretarial studies courses have been renamed 'Office Studies', so that men as well as women many acquire these and other office skills without the association with 'women's work' (and therefore presumably without embarrassment). Word processing—the activity and the technology—is shifting its domain, and the boundaries of the old gendered division of labour in the office are being redrawn.

Is this loss of a monopoly of expertise significant for women? Skilled male workers guard such monopolies jealously, for to lose them means a significant loss of occupational power and status (Penn 1982; Cockburn 1985). This is not the case for women office workers. Their traditional monopoly of expertise and skill in handling office technologies has not afforded them equivalent levels of power and status at work. On the contrary, as we have seen, office work and the women who perform it have been ghettoised and degraded. Indeed, the very concept of 'skill' has sat uneasily with women's work, precisely because of the cultural devaluation of the latter.

As word processing loses its exclusive association with this highly gendered and consequently degraded set of competencies, where does this leave the women who traditionally performed this work? In the short term, it is clear that little has changed in the way that the majority of offices are run. Despite the fact that some men can now touch-type, and no longer feel compelled to maintain their ignorance of how to operate office technologies like word processors, the kinds of activities to which they are applying their new-found skills are on the whole very different from the ones carried out by the traditional secretary or typist. While they may now use their machines to enter text or to operate other types of business software, such as spreadsheets, they tend to perform these tasks in support of their primary activities rather than as wholesale, intensive functions in their own right. That is to say, professional men are taking on only *some* elements of the text production task. There are still very few men whose entire jobs are concerned with processing other people's words, as secretaries or typists. And, just as men are not becoming full-time typists simply because they can now operate keyboards, so women are not automatically moving into the computing professions just because they are becoming computer-literate. Their technological competence in the office has not resulted in them being given more creative and responsible office tasks, or greater status. Typists are very likely to be simply left with the performance of a set of more piecemeal and disjointed word processing activities than before, for example, entering corrections to drafts or tidying up the layout of documents.

In the long term, it seems that we need to develop new arenas for women office workers to move into. But more than technological change is required to change patterns of working and the traditional positions of workers. Organisational and political changes in the workplace are also necessary. These are much more complex, conscious and long-term processes. As Huggett (1988) has suggested, the spread of microcomputers in the office provides secretarial workers with *potential* opportunities to claim responsibility for a range of activities (for example, desk-top publishing or database management) which would broaden their jobs and expertise. And there have been initiatives, in Britain, in Scandinavia and in the Australian Tax Office (ATO), to broaden the remit of clerical workers more generally, and to allow them to develop a range of computer-related skills such as systems support or systems administration. However, this requires a recognition of women's competencies and abilities, rather than a denigration of them. It requires a recognition that full-scale political and organisational changes in the conduct of offices are necessary. New technologies may provide loopholes and opportunities for change, but they do not of themselves create it.

Conclusion

Both office automation and the automated office have mutually conditioned one another and have taken a highly varied form. The logic behind information technology equipment design has not simply reflected technical or even economic rationality, but has also reflected the gendered nature of the work for which it has been designed. Office technologies have been fluid in form, undergoing fundamental changes over the last couple of decades. Moreover, once established, these systems have been configured in various ways which reflect local office contexts and requirements as much as hardware and software exigencies, and this has had crucial impacts on the design of subsequent information technology generations. The implementation of information technology in the office, then, has involved the fusion of technical and social/political elements such as work organisation, managerial practices, gender relations and expertise which are in mutual interaction. This suggests that the relationship between technologies, patterns of work organisation, and social relations in the workplace is not a simple one, but involves mutual shaping and reshaping, with technologies emerging in particular social contexts in response to particular problems, influencing these and themselves being reshaped in response to the changing patterns of work organisation and social relations within which they are situated.

This kind of perspective is important not only because it furthers our theoretical understanding of the technology/society relationship. It has practical, political implications too. An appreciation of the various inputs into technologies shows us that we have some control over the design and configuration of these technologies, allowing us to intervene and to be more proactive in technological and organisational development.

This is clearly of great importance to all those of us concerned in one way or another with the improvement of working conditions or the advancement of the labour force. Technological options are not as closed as they sometimes appear to be, and we may set about designing technological systems which promote our own political objectives, such as the broadening of jobs or the utilisation of workers' skills. Already initiatives are being taken in various countries to develop office systems and work structures which extend and enskill rather than diminishing and deskilling. Some of these concentrate on ensuring the participation of office workers in new system development so that the system design will promote the use of their skills and abilities. Other projects concentrate more on office job design, attempting to create new structures and job categories which are based on the recognition of women's skills. In my view, the best initiatives are those which address both systems and jobs, seeing neither of these as beyond their prerogative. Such initiatives promote broader recognition of the social shaping of both technologies and jobs, and the ways in which the two are inter-related. Having rejected the view that these matters are beyond our control, we can begin to develop strategies for technological and human development.

Notes

[1] I am most grateful to Mary Jennings; her insights and expertise contributed greatly to the development of some of the ideas of this paper. This paper draws upon research funded by the Science and Engineering Research Council, and by the Economic and Social Research Council's Programme on Information and Communications Technologies (PICT).

4

New technology and work organisation

The role of gender relations

EILEEN APPELBAUM

This paper examines the role that gender relations play in shaping the transformation of production systems currently underway in the industrialised economies. Firms are experimenting with innovations in management methods and work organisation as educational levels of workers rise and information technologies gradually supplant electromechanical automation as the dominant production technology. In the United States, for example, computers and related information-processing technologies account for nearly half of new spending on capital equipment (OTA 1990). Information technology's share of the overall capital stock of US corporations has doubled to 13 per cent since the 1970s, and now exceeds 25 per cent of the total in communications, finance and insurance, business services, health, and education—all large employers of women (Roach 1988). Increased competition is challenging firms to transform management methods and work organisation to take advantage of the capabilities for diversified quality production inherent in these new technologies.

Indeed, it is something of a puzzle why the opportunities created by information technologies have not had a more dramatic effect on management and work methods. In part, the answer is that previously established institutional structures and old forms of work organisation continue to exert an important influence as new technologies are introduced. Gender relations figure prominently among the organising principles that shape the implementation of new technology, though the stratification of the labour force by gender is too often neglected in analyses of organisational change (Wajcman 1991, Ch. 2). Yet gender relations in employment—the sex stratification of industries and occupations, the payment of lower wages in industries and occupations stereotyped as female, the over-representation of women in part-time, temporary and other contingent work arrangements, the unequal access of women to company-sponsored training, the reliance by firms on high turnover to mask the use of women as a labour reserve—have an important influence on the strategic decisions of firms and unions as these new technologies are adopted.

The analytical argument, to be developed in subsequent sections of the paper, can be briefly summarised as follows: the implementation of new information technologies entails a paradigmatic shift in the production processes of firms that opens up a wide range of alternatives for the development and utilisation of worker skills. While management has always had some discretion in determining how to deploy technology and organise work, information technologies are far more malleable and apply more broadly across industries and occupations than earlier generations of automation. Numerous studies (see, for example, Appelbaum 1985; Zuboff 1988; Appelbaum & Albin 1989; Kelley 1989; Merchiers 1991) have shown that the application of information technology and the reorganisation of work are not driven exclusively by technological factors. To the contrary, information technology has dramatically increased the range of choices available to managers, whose decisions about the implementation of specific technologies are mediated by social institutions and strategic considerations.

Institutions act as both opportunities and constraints, ruling out some types of behaviour by management and encouraging others, and contributing to nationally distinctive patterns in organising production. Widespread transformation of production into high performance systems is by no means inevitable. According to the German industrial sociologist, Wolfgang Streeck, 'neither technological change nor the evolution of product markets are by themselves sufficient to move industries into a diversified quality production mode' (1991, p. 32). What is required are social and political institutions that impose constraints on rational actors 'so as to protect profit-seeking individuals from the temptations of hyper-rationality and prevent them from seeking short-term, quick-fix solutions . . .' (p. 49). Otherwise, firms in competitive markets will have difficulty resisting the temptation to exploit any weakness in the market power of the workforce for short-term advantage, and difficulty justifying the expenditures on investments in the skills and capabilities of front line (production) workers that transformed production systems require.

In industries such as communications, banking, insurance, and business services—where front-line jobs stereotyped as female and compensated accordingly are directly affected by information technologies—managers following the logic of individual profit maximisation will tend to make choices that are economically sub-optimal for the society as a whole. That is, to the extent that gender relations influence human resource policies, and female employees are in an inferior labour market position than are male, managers will be led astray when making decisions affecting the efficient organisation of production by the lower wages and more precarious employment status of women workers and by their own prejudices. If firms fail to make adequate investments in the skills and job security of female front-line workers, however, firm performance on a variety of quality and efficiency measures will suffer in comparison with state-of-the-art benchmarks, despite large investments in hardware and software.

Moreover, it should be noted that these industries—which figure prominently both as employers of female workers and as investors in information technologies—supply producer services to firms in virtually every industry in the economy. Improvements in production methods here can have positive spillover effects on the cost or quality of purchased inputs in every other sector of the economy. Thus, the development of an institutional framework that rules out deskilling alternatives in the application of information technology to women's jobs has important effects not only for the women affected but for economy-wide improvements in productivity as well.

A new technological paradigm

A decisive shift in the dominant technological paradigm organising production is occurring in the industrialised countries. The use of mass production technologies, rigid hierarchies, and the detailed division of labour to produce standardised products that compete in price-conscious markets is giving way in an increasing number of industries to the use of information technologies to produce diversified goods and services that compete on the basis of quality. Information technology—the combination of microprocessor, computer, and telecommunications technologies—has become pervasive, extending to nearly all sectors of the economy in industrialised nations, and affecting the jobs of large numbers of workers. Office automation and inventory control technologies have profoundly altered production in industries largely unaffected by earlier rounds of technological innovation. Accounts of 'paper-less' offices, 'operator-less' telephone systems, and 'teller-less' banks figure prominently in descriptions of the new efficiencies to be achieved with these technologies; though the fact that it is the pink-collar clerical, administrative, sales and service jobs overwhelmingly filled by women that are on the cutting edge of these changes has gone largely unexamined.

Information technologies have made possible a range of new products and services, and have altered the costs of production in services as well as goods. In contrast to mass production technologies, in which highly specialised equipment dedicated to the production of a single component or product achieves low unit production costs by turning out large numbers of identical products in long production runs, information technologies make small batch production cost competitive with high volume, and allow equipment to be easily reprogrammed to produce highly diversified outputs. As a result, specialised high-quality goods and services to meet the requirements of small markets and even individual customers can be produced cheaply. Small firms are able to achieve unit costs that previously required large scale. Alternatively, computerised production equipment that can be easily and quickly reconfigured to manufacture a variety of related products has made it possible for large firms to achieve high volume production while customising high-quality products or by combining standardised components into highly differentiated final products to meet the demand for diversity in mass markets. Thus, information technology has opened up new *possibilities* for flexible specialisation in smaller firms as well as for flexible automation in high volume producers. *Realising* the full potential of information technologies for flexibility, however, is not merely an engineering exercise—a question of designing equipment with the correct specifications. The shift from electro-mechanical mass production technologies to microprocessor-based information technologies also entails important design choices regarding management methods, work organisation, and industrial relations that affect the flexibility of the new technologies.

Evidence is slowly accumulating that a less rigid bureaucratic firm organisation and a less polarised distribution of job skills improves performance along many dimensions, including cost, productivity, quality, timeliness, and responsiveness to customer needs. Yet such changes are proceeding slowly in the US despite increased pressures for innovations in management and work organisation as a result of sharply rising competition in world and domestic markets, and despite the apparent success of markedly different management practices in some countries whose firms compete with US producers.

Managerial discretion in the implementation of new technology

A striking feature of microprocessor-based technologies is the wide latitude for choice that managers have with respect to how specific technologies will be implemented. The malleability of these technologies is apparent in cross-country comparisons of their deployment in firms in a given industry, and even in different firms in a single country. Whatever the industry, it is not possible to view the skill requirements and organisation of work associated with computer and communications technologies as determined by the engineering specifications of the hardware. Indeed, nationally distinct patterns appear to be emerging.

A major study of factory automation in Germany and the United Kingdom found a 'striking variety' in the organisation of the work process under computerised numerical control, and in training and skill patterns (Sorge & Warner 1986, p. 159). In Germany, the application of computer numerical control (CNC) machines is governed by the large proportion of highly skilled workers in the labour force and by a blurring of distinctions between the workshop and the office, craft workers and technicians, shop floor workers and supervisors. The opposite is true in Britain (p. 163).

Similar conclusions emerge from recent comparisons of French and Japanese factory organisation. In France, the work of machine operators is increasingly mechanised, and sharp distinctions between operators and technicians and between technicians and engineers are maintained. The technician category has become more important and replaces the promotion of skilled operatives. The result is an increasing polarisation of skills. Occupational categories are very different in Japan. Beginning engineers are assigned to the technician category and can also be given the tasks of workers before they move on to engineering. The emphasis is on the simplification of the production system, on quality control, and on the upgrading of worker skills through the overlapping of functions and tasks and the flexible definition of assignments (Merchiers 1991).

Similar differences in the design and reorganisation of work around computer and information technologies emerge in comparisons of US industry with that of Japan or Western Europe. The Massachusetts Institute of Technology (MIT) Commission on Industrial Productivity concluded that, in comparison with its most important international competitors, US industrial performance has suffered from management's failure to co-ordinate design and manufacturing functions, from management's neglect of process design and production operations, and from management's low regard for training and the failure of firms to invest in worker skills (Dertouzos, Lester & Solow 1989, pp. 70, 72, 77, 82 and 99). Continued conflicts over the rights of workers to unionise and to participate in enterprise decision making has produced an adversarial pattern of industrial relations with little trust between workers and managers.

Shoshana Zuboff (1988), in her influential study of work organisation and labour skills, identified the problem as one of managerial ideology. Her case studies illustrate that the way in which new technologies are implemented depends on whether US managers are willing to relinquish their centralised control over knowledge and authority. Information technology, she demonstrates, can be used to *automate* operations 'according to a logic that hardly differs from that of the nineteenth century machine system' or it can be used to *informate* the processes to which it is applied. In the latter case, 'it increases the explicit information content of tasks and sets into motion a series of dynamics that will ultimately reconfigure the nature of work and the social relationships that organise productive activity' (pp. 9, 10). Examining the implementation of computer rationalised production processes in the insurance industry, I reached the same conclusion:

> Two distinct responses of firms to problems of control over computer rationalised production systems have emerged . . . The goal of an algorithmic design approach is to limit workers' skills and to reduce the role of human knowledge and judgment in the production process by designing ever more perfect, self-regulating production systems . . . [A] second path of development . . . [uses] low cost microprocessors . . . to increase the knowledge and expand the capabilities of a wide range of workers (Appelbaum 1985).

Maryellen Kelley in her work on programmable automation (PA) in machine tools also concludes 'that there is no technological imperative shaping the division of labour' in US industry (1989, p. 246). She distinguishes among three patterns of application of PA technology, with very different implications for the skills of blue-collar workers and for the extent of polarisation of skills.

As these studies of information technology and work organisation show, there is neither a technologically determined outcome nor a single management policy driving the reorganisation of work around these new technologies. Whatever the application, there are always choices that can, and are, being made. Moreover, technology not only influences the way in which work is organised and skills are generated and defined, but depends, in turn, on these institutional subsystems as well as on the commercial strategies of firms (Sorge & Streeck, 1987). In addition to altering production processes, information technologies have led to the development of new products and services that blur old distinctions among industrial sectors and allow firms to redefine the markets in which they compete and to reconceptualise their lines of business (Baran & Gold 1988, pp. 14, 19). Thus the direction and substance of technological change—whether it will be used to transform production into high performance systems or will be used, instead, to deskill workers and increase monitoring—is mediated by social institutions and firms' strategic considerations.

Institutions as constraints and opportunities

Market forces alone are unlikely to generate transformed production systems. Firms engaged in the high volume production of customised quality-competitive goods and services or flexibly specialised to meet changing needs in particular market niches need to invest in the *redundant capabilities* of their workforce, and in *collaborative* rather than competitive relationships with other firms in their networks. Flexibility in responding to changes in market conditions and opportunities requires that firms invest 'in general capabilities for as yet undefined future purposes [but this] is not easy to justify for managements under competitive pressure' (Streeck 1991, p. 36).

Streeck has identified several types of redundant capabilities required by firms with transformed production systems. High performance organisations need workers with high-level general skills who are able to move flexibly through a wide range of production, inspection, maintenance, and administrative tasks. Workers in such settings have more skills than are required at any one time—that is, they have redundant capabilities. Duplication and overlap of functions is a redundant capability that becomes necessary as the boundaries between research, engineering, and marketing begin to blur. This facilitates co-operation among departments, but it also means that firms have capacities that, while they are expected to prove useful at particular junctures, usually remain unused in the background. Decentralised decision making also necessitates redundant capabilities because it requires information and authority to be shared among people at different levels of the organisation, all of whom must be competent to make decisions as the opportunity arises. Finally, a high level of mutual trust and loyalty between employers and workers, predicated among other things on long-term employment commitments, is a necessary input in diversified quality production.

Rational managers in hierarchical firms lack the incentives to make the large investments required in training front-line workers, decentralising competence, moving decision making down the hierarchy, reorganising work, and restructuring the employment relationship. Some analysts have gone further and identified positive incentives for 'management opportunism'—that is, for decisions by managers that sacrifice efficiency in order to protect the narrow self-interest of managers and the perceived benefits of managerial control over the work process (Smith 1991). The incentive structure in competitive markets similarly prevents rational, profit-maximising market participants from developing strategic alliances and collaborative relationships with firms that, in other contexts, may be their competitors. Yet such collaborative networks among firms ensure the quality of finished products and facilitate a quicker response to changing market conditions.

There is, thus, a failure of markets and hierarchies. Rational decision making by managers in profit-maximising firms is likely to lead to outcomes that are sub-optimal from the point of view of the requirements of diversified quality production and high performance systems. This is easiest to see in the case of investments in training, where rational calculations of the firm's ability to capture the returns to such investment leads to an overemphasis on providing workers with narrow, firm-specific skills. But it applies as well to other redundant capabilities. For example, delegation of authority and decentralisation of information and competence threaten the hierarchical management, authority, and command relations typical of most firms; and management may, for this reason alone, prefer less rather than more decentralised decision making. Or, to take still another example, retaining workers when sales decline during a downturn in the business cycle in order to reap the benefits of mutual trust over the long run imposes high short-run costs. Individual firms that choose to incur such costs may find themselves undercut by competitors who underprice them in the short run and drive them out of the market before the gains from greater trust can be realised.

What is needed to overcome this failure of markets and hierarchies is the development of social and political institutions that constrain self-interested behaviour on the part of firms or managers in order to reach outcomes that, once achieved, are superior both for the economy as a whole and from the perspective of the individual firm. Examples of such institutions, all of which can be found in one or another of the industrialised economies and none of which exists in the US, include legal protections against redundancy that promote long-term employment relationships, joint public–private work-based training consortia that overcome the reluctance of firms to invest in broad skills, wage norms that prevent the lowest wages from falling below half the average wage in manufacturing, associations of small firms that negotiate technology transfers from larger firms with which association members regularly do business, and codetermination laws that mandate that management share information and decision-making authority with workers. Alternatively, strong unions and a flourishing labour movement can co-ordinate wage bargaining and establish wage norms that eliminate competition on the basis of very low wages, can negotiate joint labour–management participation in decisions regarding technology, training, and health and safety issues, can negotiate employment security, and so on.

In an institutional framework, dense with social and political institutions that rule out the worst excesses of predatory capitalism and essentially take wages out of competition, firms are denied the ability to drive down wages and are constrained to find alternatives for remaining competitive in world markets. Profit maximisation subject to these constraints leads managers to make difficult decisions regarding changes in management methods and work organisation required to achieve continuous improvements in productivity and quality via the transformation of production and to produce for quality conscious markets. Thus, constraints on behaviour may simultaneously provide positive incentives and opportunities for the development of high-performance production systems.

One might point to Japan and the former West Germany, economies with very different but nevertheless very dense institutional structures, as exemplars of these changes. The older, 'Fordist' production model is being replaced by production systems in the core industries of these economies based respectively on lean production and on diversified quality prduction. Both economies have solid records of achievement on such indicators of competitiveness as overall productivity gains and net exports, despite steady increases in wages; manufacturing wages in Germany (measured on a market exchange rate basis) now exceed those in the US and in Japan have nearly drawn even. But these descriptions of German and Japanese success ignore the peculiar role played by gender relations in employment in these countries. Some opportunities for wrong choices with respect to work organisation are ruled out by the simple expedient of social norms that exclude large numbers of women from participation in employment—both countries have very low labour force participation rates for married women—and segmentation mechanisms that discourage women from pursuing university degrees and limit their access to professional and managerial occupations and careers.

In contrast, decision making by US firms is mainly market driven, and social and political constraints on management are largely absent. The announced goal of the National Association of Manufacturers—a union-free America—is in danger of being achieved. Union representation in the private sector, at twelve per cent of private sector workers, is precariously low. Other constraints on firm behaviour are either nonexistent or nonbinding. The legally mandated minimum wage, for example, has fallen from half the manufacturing wage in the 1960s to less than a third today. At the same time, female labour force participation rates in the US are high, and continue to rise. Women in the US provide both a pool of low wage workers still sharply stratified by occupation and a pool of highly educated workers (22.3 per cent of female workers have a university degree compared with 24.5 per cent of male workers) still earning substantially lower wages than their male counterparts and still facing substantial occupational barriers.

If social institutions mediate choices regarding the implementation of information technology, then one should expect these gender relations in employment to shape technological choices in the US The availability of a female work force with these characteristics can be expected to influence management prerogatives in organising production.[1]

The role of gender relations

An important characteristic of computer technologies, especially in their application in white-collar industries such as banking, insurance, and communications which are major employers of women, is that they eliminate the *least* skilled clerical, managerial and professional jobs *regardless* of how they are implemented.

In an early and important study of sex stratification, technology and organisational change in the telephone industry, Sally Hacker (1979) found that women held the lowest paid positions in both management and non-management occupations at AT&T, and that these were precisely the jobs displaced by technology. She suggested that the increase in skill levels as industry moves from mass to process production technologies 'merely reflects the elimination of women's work from the occupational structure' (p. 54).

In insurance, skilled clerical and more routine professional jobs that were the rungs on a career ladder from clerical to professional jobs disappeared in the US in the 1980s. They had provided internal mobility for young men without formal professional qualifications and for young women as well, after favourable affirmative action rulings in the 1970s. Jobs such as rating policies to determine premiums or assessing risk and underwriting individual life or automobile insurance have been eliminated right along with routine clerical and mail room jobs (Appelbaum 1987). Similar effects of technology in the financial sector have been observed in Europe as well (Bertrand & Noyelle 1988). And a study of financial services in Australia found that '[w]hile there has been a distinct feminisation of the insurance sector over the last two decades, this has also been associated with the collapse of the traditional career structure' (Stevenson & Lepani 1991).

A comparative case study of insurance, banking, and telecommunications services in the US found that both operators and routine clerical workers 'are experiencing the elimination rather than transformation of their functions' (Baran & Gold 1988, p. 69).

As a result of the elimination of routine management and non-management jobs held by women in these industries, an increase in skill levels has generally been associated with the implementation of information technologies. This tells us little, however, about whether firms, in Zuboff's terms, have chosen to informate or merely to automate. The interesting questions have to do with the effects of new technology on the organisation of work and on the skill requirements in the jobs that remain. Here, gender relations play an important role.

Low wages, part-time work and technological choices

The observation that technology has displaced many routine clerical jobs in industries like insurance has given rise to the view that office automation is upgrading the skill requirements of the jobs women hold. Baran and Parsons, writing in 1986, argued that while the early stages of computerisation tended to automate work in conformity with the Taylorist fragmentation of tasks and bureaucratic administration of work typical of Fordist work organisation, the logic of the later stages of office automation is very different. They write (p. 70):

> The emerging organisation of work bears little resemblance to the assembly-line. Whereas the ideal operative in Ford's factory turned one bolt, the ideal production worker in the insurance company of the future with the aid of a sophisticated computerised work station, rates, underwrites, and issues all new policies for some sub-set of the company's customers, handles the updates and renewals on those policies, and as a byproduct enters the information necessary for the automatic generation of management reports, actuarial decisions, and so on . . .

> In the process, much routine clerical activity is simply being eliminated; the huge data processing sweatshops are being closed as companies move to single-source entry and attempt to push the entry function as close to the point of data generation as possible . . .

A study of the Norwegian National Insurance Administration, in which the introduction of an on-line processing system was a strongly negotiated process between management and trade unions representing clerks and officials, found that while total employment in 1988 was about the same as in 1973 (though below its 1984 peak), the composition of employment had shifted from 63 per cent clerks, 24 per cent officials and 13 per cent managers to 12 per cent clerks, 77 per cent officials and 11 per cent managers (Byrkjeflot & Myklebust 1991, p. 61). The authors conclude, however, that in this instance the results would have been the same if the process had been spontaneous rather than negotiated, and that the results apply more generally to the computerisation of certain types of office work. The negotiated decision, write the authors,

> was based on the assumption that the new system would actually lead to a reduction in office work (e.g. all filing and all accountancy work would be performed automatically by the system). Thus the result of the negotiations can partly be regarded as the result of a process of computerisation even if rational, interest-oriented actors decided the matter. For what other kinds of decisions could they have taken? The process of computerisation described above seems typical for service-producing organisations whose prime tasks are performed by a tier of semi-professional employees supported by a tier of clerks . . . (p. 62).

Current developments in work organisation in the US insurance industry contradict these optimistic expectations and indicate that there are, indeed, other kinds of decisions that can be taken. With less than 3 per cent of the labour force in this industry unionised (Hartmann & Spalter-Roth 1989), it is possible to observe the results of a (mostly) spontaneous process. The 1986 predictions of Baran and Parsons have generally not come true. While a few smaller personal lines companies and the highly specialised carriers that insure non-standard commercial risks (elevators, cargo ships, etc.) do approximate the Baron and Parsons description, the large companies insuring standardised personal and commercial risks (health, life, automobile, homeowner's, commercial automobile or buildings, worker's compensation) and employing most of the insurance industry's labour force continue to organise data entry and other routine clerical functions along Taylorist lines.

In the late 1970s and early 1980s insurance carriers used newly available telecommunications technology to remove data entry jobs from the central business districts of cities such as New York, Newark, Philadelphia, and Hartford, where they had provided full-time employment at full-time wages and benefits for a largely black, female labour force of primary income earners. These women were left behind in the new division of labour made possible by technology, and faced high rates of unemployment or re-employment in casual jobs at lower wages in retail trade or service occupations that are the common lot of minority workers in the inner cities of the US.

Now, a decade or more later, these insurance companies still utilise large data entry campuses located in middle-class suburbs or small, economically depressed cities where white women, even those with some formal education beyond high school, have few alternative employment opportunities. In this way, they are able to take advantage of the keyboarding skills and attention to detail of a female labour force that, lacking local employment alternatives, can be hired at wages that are low in relation to educational levels, often on part-time hours, to do routinised data entry, which the women are nevertheless motivated to perform carefully and well.

Typically in these operations, policy applications sent in by agents are screened; routine risks are then entered into the computer by a clerk, while non-routine applications are sent on to an exceptions handler. The assessment of risk and rating are done by the computer, which also produces and issues the policy. As a result, as Baran and Gold point out, 'functions formerly divided among entry clerks, raters, underwriting assistants, and underwriters themselves, have been consolidated . . .' (1988, p. 64). The jobs do involve high levels of individual responsibility for accuracy and quality control, but they are still essentially dead-end data entry jobs requiring some keyboarding and problem-solving skills, but physically separated from professional jobs and from direct contact with administrators or clients, and with no possibility of career development. Baran and Gold's characterisation of the woman holding this job as 'a highly skilled clerical worker' does not seem justified.

Possibilities within this work organisation for further productivity gains and labour cost savings—beyond the initial improvements associated with routinisation and computerisation of data entry and the payroll reductions associated with replacing full-time, primary income earners with a part-time labour force—are not available. Indeed, lacking the means to achieve such gains, some US firms have begun moving these routinised clerical operations off shore to lower wage countries like Jamaica, Barbados, and Ireland in order to achieve further reductions in labour costs.

Thus deskilling of clerical work, abetted by investments in computer and communication technology, remains very much an option as a means of rationalising a significant part of the production process in insurance—especially back-office work carried out by a still-large and overwhelmingly female labour force with less than a four-year college or university degree.

Despite the potential for integrating tasks and reducing levels of management and supervision, rigid, bureaucratic structures, a functional division of labour and the separation of conception from execution survive to a surprising extent in these organisations. There have been positive changes, some associated as much with earlier affirmative action decisions as with new technology. Many companies have created new, multi-skilled customer service representative jobs, utilising information technology to deal immediately with customer inquiries or complaints about policies or claims, without referring these requests to professional employees. These jobs are classified as clerical, and remunerated accordingly, leading to complaints about 'upskilling' and 'downwaging'. In addition, the collapse of traditional internal career paths has been accompanied by increased demand for credentialled insurance industry professionals hired directly through the external market. In the US, there has been a significant movement of college-educated women into these jobs.

In terms of the argument regarding the role of institutions developed in the previous section of this paper, the ability to hire relatively poorly paid female workers hampers managers in industries like insurance and banking in their efforts to justify (to themselves as much as to the upper echelons of their organisations) the investments in redundant capabilities required by a high-performance production system. Having data entered as it arises and restructuring clerical jobs so that they are full time or, alternatively, are able to provide a permanent, part-time work force with conditions of work comparable to those of full-time workers (pro-rated health and pension benefits, wages that rise with seniority, access to training) implies a substantial increase in payroll costs. Redesigning jobs to integrate data entry with other, higher level tasks requires investments in training for a work force for whom employer-sponsored training has typically been measured in hours. These are also jobs which, in view of their fragmented, repetitive, dead-end nature, have been associated with high turnover; an outcome which is then used to justify the firm's reluctance to invest in the workers who hold these jobs and makes any transformation of the production process more difficult to undertake. The reorganisation of production is inordinately costly under these conditions—it involves, first, the costs of bringing a contingent workforce up to the conditions of work enjoyed by regular workers, and then the costs of investing in the redesign of jobs and the training of workers.

Yet continued reliance on Taylorist organisation of production is also costly, though the costs may be less evident. In addition to the costs incurred in continuing to maintain numerous layers of bureaucracy, which better use of technology now makes redundant, Taylorist organisations effectively rule out continuous quality improvements and cumulative productivity gains in a large part of the production process in office industries. The extent to which this is related to the so-called productivity paradox—the zero, and even negative, productivity gains in insurance and other financial services in the US during a decade in which massive investments in computer and communications hardware were made—remains a matter of speculation. But would anyone suggest that there is no association? In any event, the inability to achieve further productivity gains in these industries has provided a strong incentive for moving back-office work off-shore, reducing job opportunities for women in particular local labour markets, and imposing significant social costs.

Women workers and the myth of attrition

Information technology, unlike earlier generations of automation, has had an enormous impact on occupations typically held by women—telephone operators, clerical operatives, lower level managers and professionals. As noted above, many of these functions have been eliminated by automation, and the women holding these jobs displaced by technology. Yet, there has been little research on the technological displacement of women workers. Little is known about the problems of adjustment faced by these women, or about the conditions of work and pay under which women, especially those lacking a four-year college or university degree, find re-employment. In part, this phenomenon of technological displacement has been masked in the US by the steady rise in women's employment and the continued growth of clerical jobs. The effects of technology can be seen in slower growth of clerical work, rather than its absolute decline, with the result that clerical work fell as a proportion of white collar employment in manufacturing, finance and insurance, and services during the 1980s (Attewell 1990). In part, the high turnover of women in many of these jobs has meant that the restructuring of employment has taken place without the necessity of massive layoffs.

The implicit assumption is that the process has been benign. Women, trying to reconcile the demands of family and work, move in and out of employment voluntarily. The relatively high turnover of women in occupations such as telephone operator or clerical worker allows individual firms to use attrition to restructure employment humanely in response to changes in technology or markets. Women re-entering employment have no expectation of returning to their old jobs, and would have to find new jobs in any case. These new jobs are assumed to be at the same skill and wage level as the original job, except to the extent that the woman's skills may have deteriorated or become obsolete during her absence. A reassuring story all the way around. But is it true? Evidence from the telephone industry is disquieting.

Sally Hacker (1979), in her study of organisational change in AT&T during the 1970s, paints a less reassuring picture. She suggests that women were manipulated into functioning as a flexible source of labour (or labour reserve) in order to facilitate changes in technology, deliberately drawn into formerly male jobs as they were routinised in preparation for elimination by automation (sometimes in the guise of affirmative action), then displaced by technology under cover of high turnover and attrition.

Employment at AT&T shifted in the post-war period from 70 per cent female in 1946 to 58 per cent in 1960. Total employment at the telephone company increased, but there were actual decreases in the employment of women as operators and in lower level clerical positions. Women held most of the 80 000 jobs eliminated at AT&T between 1958 and 1964. Data from one of the AT&T operating companies providing local telephone service shows that during the 1960s thirty-eight job titles were eliminated, of which thirty-one (primarily clerical) were traditionally female jobs. Not just the company, but union officials as well, saw no problem with any of this because the jobs affected by technology were mainly held by telephone operators and clerical operatives whose turnover was high. The union acquiesced in the displacement of thousands of union workers because the high turnover in these occupations meant it could be accomplished without layoffs of union members and because the overall expansion of employment in the industry meant a steady increase in union membership during this period despite the loss of these women workers. The union grew from less than 600 000 members in 1947 to a peak of nearly 1.1 million in 1982. The women were dispensable. Certainly, the company had no wish to increase its retention of women either by making these jobs more attractive or by training women for the expanding opportunities in men's jobs, and the union did not push for such changes. High turnover was accepted as an inherent characteristic of women workers.

The women themselves were less sanguine, and the National Organisation for Women took their complaint to the US Equal Employment Opportunity Commission, which in 1971 called AT&T the largest discriminator against women in the US. Affirmative action remedies were imposed on the company, and the effects of these on women's employment between 1973 and 1976 were studied by Hacker. She found that over 36 000 women's jobs were eliminated during the three-year period—mainly operators, low-level clerical workers, and lower level supervisors. A very small proportion of women and minority men with higher levels of education or technical skills were advantaged by affirmative action, which opened middle management positions and some new positions created by technology to them. Women were also encouraged to enter certain 'male' craft jobs—especially as frame technicians and telephone installation and repair technicians. But planned technological change eliminated more women's jobs during this period than affirmative action provided. Hacker notes something suspicious about the craft positions women were encouraged to enter: these were precisely the craft jobs slated for elimination or reductions as a result of technological change. Moving women into jobs ultimately slated for displacement by technology after they were first simplified and routinised could serve several purposes, including letting attrition take its toll. While Hacker had no direct evidence that management at AT&T was deliberately using affirmative action to move women into previously all-male jobs in order to prepare for the displacement of workers by computer and fibre optic technologies, she was able to cite evidence from the Bureau of Labor Statistics showing that 'in four Bell Companies [local telephone service companies], AT&T increased the workload and reduced flexibility on the job to help maximise attrition [by women] during and after a shift to less labour-intensive telephone technology' (1979, p. 551). Thus, the corporation was certainly capable of manipulating attrition to avoid the necessity of layoffs.

This view of attrition as humane continues right up to the present time. A very interesting and well executed study of the effects of technology on employment at a local telephone operating company between 1980 and 1985 (Lynch & Osterman 1989) describes the company as 'benevolent' because it relies on attrition and incentives for early retirement to reduce employment. The most significant adjustment mechanism at this company is transfers to other jobs, which accounted for 50 per cent of all outflows from particular occupations.

Employment at this operating company decreased by more than 13 000 workers over the five-year period. Technological change increased employment in some occupations, but adversely affected operator services (still overwhelmingly female despite the entrance of men into this occupation following the earlier affirmative action ruling) and network distribution occupations, which included the craft jobs which women had made some progress entering. Results are not reported by gender for each occupation, but the study found that transfers accounted for 70 per cent of departures in network distribution (still mostly male) but only 36 per cent of those in operator services (still overwhelmingly female). The study provides evidence (p. 203) for the

> often-stated, but rarely demonstrated observation that firms use high female turnover rates as an aid in adjusting to employment shifts. Despite the fact that women are heavily represented in the most seriously impacted occupation [operators], the Company did not have to rely exclusively on layoffs. Furthermore, the adjustment did not occur via unusually high transfer rates since the equations show that occupations with large fractions of women have lower transfer rates relative to exits than do other occupations. High female turnover, not layoffs or transfers, constituted the adjustment mechanism . . .

Doubts about the benign nature of turnover as an adjustment mechanism in this industry are fuelled by data on female earnings. Among the private sector service-producing industries that are major employers of women—telecommunications, retail trade, finance and insurance, business and repair, personal, medical, social, and miscellaneous—only telecommunications is highly unionised. Women's average weekly earnings are highest in this industry—exceeding even the average earnings of men in most service industries (Hartmann & Spalter-Roth 1989 p. 5). Women displaced by technology from jobs in this industry are unlikely to replicate the pay and benefits in other employment.

The union, having lost members after 1982 as overall employment in traditional telephone services declined, is no longer sanguine about the loss of employment by women. New strategies for retaining these women both as active employees of the telephone companies, whose non-unionised information and communications activities continue to grow, and as active union members suggests that earlier union acquiescence in their termination represented a failure of empathy and imagination.

Access to training

In general, there is a tendency in the US for profit-maximising firms to under-invest in training. Worker skills, especially the broad skills needed by firms that adopt more flexible work organisations, are difficult for the firm providing the training to appropriate. Trained workers can be hired away by other firms that fail to provide training and, instead, purchase the skills they need by hiring workers who have been trained by other employers. Thus, firms cannot be certain of recouping investments in their workers, and may find that their investment in training has simply increased the pool of skills available to their competitors. Under these circumstances, firms will be tempted to restrict the amount of training they provide to their workers, and to limit training to job- or firm-specific skills that are not easily transferable (Finegold & Soskice 1988). High-performance work organisations rely on worker skills to increase the range and quality of products and services they provide and to adapt swiftly to later generations of technology, but there are strong incentives for the level of investment in such skills by profit-maximising firms to be sub-optimal.

The problems associated with justifying expenditures on training become especially acute in the case of women in front-line clerical, sales or service jobs. In the first place, there is no history of enterprise-based training for women workers. As workplace training is currently administered in the US, employers have sole responsibility for the selection of training participants and administration of training programs. Numerous studies have shown that race, gender, and amount of formal education are important determinants of who receives such training in the US. An analysis of 44 388 individuals included in the 1983 Current Population Survey conducted for the US Department of Labour found that the probability of receiving employer-sponsored training was significantly higher for men than for women, for whites than for non-whites, for those who are more highly educated, and for professionals and managers (Kochan & Osterman 1991, p. 73). The US Census Bureau's Survey of Participation in Adult Education found that 58 per cent of employer-provided training went to workers in executive, managerial, professional, and technical positions, while these workers accounted for only 27 per cent of employees (Carnevale & Goldstein 1990, p. 49).

Despite the fact that training is effective in raising wages and productivity for women and non-whites when they receive it, numerous studies have found that company-provided training is concentrated on white males (Lynch 1989; Flynn 1990; Kochan & Osterman 1991). Moreover, women are likely to receive less training even when they are included in company-sponsored programs. The duration of firm-based training for women in the National Longitudinal Samples survey of young workers was just 70 per cent of that for men (Lynch 1989, p. 28). Only the fact that women are much more likely than men to pay for their own training in community colleges or other post-secondary institutions equalises the incidence of training between men and women (Flynn 1990). But training outside the workplace is by no means as effective as work-based training in raising either productivity or wages.

Conclusion

In contrast to the electromechanical technologies that automated factories but left white-collar and clerical jobs untouched, information technologies are having as important an effect on production processes in offices as in factories. Women's jobs are fundamentally affected by these new technologies. A corollary of this development, particularly in countries like the US where a major proportion of front-line workers in white-collar industries are women, is that continuous quality improvements and cumulative productivity gains cannot be achieved without paying attention to the more efficient design of women's jobs and to reconsideration of the place of these jobs within the organisation.

Unfortunately, there is little evidence that this is happening, especially in the clerical and low-level management jobs that women hold, for the reasons we have already discussed. There is no history or tradition of attention by employers to providing enterprise-sponsored training for women, or to the integration of lower level clerical functions and lower level administrative functions into multi-skilled jobs. Nor have firms traditionally looked to jobs performed by female clerical workers and customer service representatives as a major source of continuous gains in efficiency.

In addition to the usual impediments to investment in the redundant capabilities of workers required by high-performance production systems, the transformation of work in white-collar industries is hampered by management's prior experience with women's jobs and by the disincentives built into the present system of gender relations in employment. The view of many managers is that job redesign to better integrate remaining less-skilled clerical and managerial functions is neither necessary nor desirable. Technology is expected to displace these jobs, while high turnover and attrition are expected to take care of declining employment requirements and facilitate employment restructuring. As a result, managers do not want to restructure these jobs to reduce turnover, and do not want to provide training for jobs that have a limited future. At the same time, high turnover in these jobs is used to justify the unwillingness of firms to train women workers in these or in other jobs or to plan for upgrading their skills and responsibilities. This, in turn, limits the possibilities for achieving efficiencies associated with high-performance production systems.

The alternatives—integrating the low-level clerical and managerial tasks that remain, and that are unlikely to be eliminated entirely, into multi-activity jobs; providing training to upgrade the skills of workers in these jobs—are only infrequently pursued by employers. In view of wage and benefit disparities and the possibility of relying in part on contingent workers to perform many of these jobs, the initial costs of any such reorganisation of production are likely to seem inordinately expensive in comparison with the current combination of Taylorist work organisation and gender relations. The hidden costs—in terms of missed opportunities for improving efficiency and quality—are difficult to quantify and are omitted from the usual cost-benefit analysis.

The failure to adopt high performance management methods and work organisation in the service industries that are both major users of information technologies and major employers of women in their basic production processes has economic implications that go far beyond the individual firms involved. Firms in these industries—banking, finance, business services, telecommunications—supply producer services to every other industry. Inefficiencies thus spill over into the full range of economic activities, with potentially serious consequences for overall productivity gains in the economy and for competitiveness.

Notes

[1] For a general discussion of the effects of the work force characteristics on managerial choices see Storper & Walker 1983; Bertrand & Noyelle 1988.

5

Naming women's workplace skills

Linguistics and power

CATE POYNTON

There is a real danger that workplace restructuring in Australia will exacerbate the marginalised status of many women. It is not only decisions about work organisation and the introduction of new technologies which threaten this outcome, but the failure to acknowledge the character and value of women's skills. The process of award restructuring, which presented an opportunity to redress this situation, has contributed to this neglect.

Several factors have been important. Firstly, most women in the paid workforce are clustered in a small number of predominantly female occupations. Given women's generally inferior social status this has meant that in the past not much attention has been paid to the actual skills being deployed. Further, at a time when multi-skilling has been a significant goal of workplace restructuring, the already multi-skilled nature of much women's work has received scant attention, and less credit. Indeed, restructuring has been predicated largely on men's industrial work practices which have been anything but multi-skilled—demarcation disputes being precisely about fine distinctions of who is responsible for what aspect of a job. Finally, and most significantly, women's workplace skills have always tended to the invisible, as have their domestic skills, treated as simply an extension of their identity as women: what women do because they are women. Forget about acknowledgement of multi-skilling—the first step is simply to get much of women's work acknowledged as skilled.

The What's In A Word project?[1] has been one of a number of recent projects which have intervened in the process of restructuring in order to promote greater recognition for women's workplace skills (for example Frizzell, 1991; Windsor, 1991; Cox & Leonard, 1991). However, the focus of the What's in a Word project has been significantly different, in that it has emphasised, in line with much feminist work of the last two decades, the central role of language in the social construction of 'reality', including feminine and masculine identities (Poynton, 1985; Cameron, 1985; Weedon, 1987; Threadgold, 1988).

The project began with requests from women industrial officers, working with women in the workforce (and women in human resource management in the private sector), for a new vocabulary to label a range of women's actual workplace skills so as to gain acknowledgement for these as *skills* rather than as personal attributes. Influenced by feminist thinking, they wanted to be able to 'name' women's workplace skills in ways that would enable women themselves to see a whole range of aspects of what they actually did as skilled work thus making it extremely difficult for others, particularly employers and industrial commissioners, to deny that these were indeed workplace skills. Their problem can be exemplified with one illustration: in many workplaces, women workers are required to be patient. How do you talk about patience in ways that are industrially 'respectable'?

Posing a question such as this indicates very clearly how far things have moved in considering women's workplace skills. It represents an advance even from the recent, and now widely deployed, distinction between technical, organisational and interpersonal skills developed to highlight the range of skills actually used by women (for example, Frizzell, 1991; Windsor, 1991). This set of distinctions attempted to broaden the understanding of and hence gain overdue recognition for significant areas of women's skill. However, separating technical from interpersonal skills in this way is not necessarily helpful in all cases. The What's in a Word project in effect proposed a way of defining as technical a range of 'interpersonal' skills, by seeing them as more properly language skills; that is, skills involving spoken and written language, skills which are learned and honed by practice.

Given the generally low status of 'women's skills', this is a potentially useful strategy for women in industrial contexts. For such a strategy to be effective, two things need to be established: first, that such skills are essential to the productive workplace; second, that while more women than men might demonstrate high levels of such skills, they do so because of the positions they have been socialised into occupying with respect to men and not because such skills are in any sense inherently gendered. This last is essential if language skills are to receive the recognition they deserve. To the extent that these skills are seen as the sole perquisite of women such recognition simply will not be given.

Approaching workplace skills in terms of language

What's in a Word focused on clerical work, in which a substantial proportion of the female workforce is employed. While there is still a long way to go in gaining appropriate recognition and adequate recompense for the range of technical skills deployed by women working in clerical jobs (as their technical skills have always been less highly valued than men's: see Pocock, 1988), the way to seriously bolstering women's status as multi-skilled workers has been seen to lie in gaining recognition for different categories of skill. The obvious categories are organisational and interpersonal skills.

This strategy has been problematic, partly because many of the organisational and interpersonal skills claimed for women are largely recognised only when exercised by managers. Women's claims to such skills will inevitably be opposed. A second, more difficult, issue specifically concerns interpersonal skills. Much of what women actually do in this area involves the use of language. The problems are twofold: interpersonal skills are often interpreted in psychological terms as aspects of 'personality' or personal attributes, rather than as language skills; and, there is widespread misunderstanding of and even misinformation about language, both what it is and what people do with it. As a consequence, only a limited amount of what actually happens when people use either spoken or written language is visible.

The reasons for this lack of visibility are complex. Although there is now a considerable body of research on aspects of spoken interaction, this material has on the whole not been widely disseminated outside university departments of linguistics and communication. This is partly because of the widespread resistance to claims to expert knowledge about language, particularly spoken language: every one talks and hence sees themself as an expert. As well, there are certain widespread misconceptions about the nature of language and what people do with it that serve the interests of dominant groups very well, but the interests of everyone else rather less well. People's concern with language has traditionally been overwhelmingly focused on 'rights and wrongs': 'correct' (middle-class) grammar, 'good' (middle-class) pronunciation, 'proper' (formal and written rather than spoken or colloquial) vocabulary and 'fluent' (gift of-the-gab) delivery. In this way, people are acting, intentionally or otherwise, to police one of the most effective means of maintaining separateness and hierarchy between social groups. It has also obscured the significance of language as a critical form of social action, particularly with respect to establishing and maintaining significant social relations.

Furthermore, language is understood to operate in very general or global terms, so that individuals are seen as being 'good with language', 'good with people' or 'good writers'. Such a view of language is completely consistent with psychological notions of 'personality' which represent individuals as all of a piece, singular, unitary and fixed. Such views are by no means the only available ones, however. They have been subject to substantial and cogent critique within recent feminist and post-structuralist theory (for example Belsey, 1980; Henriques et al., 1984; Weedon, 1987; de Lauretis, 1987). These writers have proposed instead that persons are not singular, unitary and fixed but that they are best understood as multiple and fragmentary at particular points in time. Such views are certainly unsettling to what has come to be the orthodox view of human personality, understood in terms of fixity and permanence. However, they are remarkably consistent with people's actual experience of functioning as different 'persons' in different situations. Within different analytical frameworks this phenomenon is understood in terms of notions such as *persona*, *role* or *subject position.* Whereas more traditional psychological and sociological notions of *persona* or *role* contain the built-in assumption that there is some 'real' or central self that remains constant, despite the adoption of various guises, feminist and post-structuralist notions of *subject position* involve no such assumption. Instead, such notions focus on the variety of 'selves' people need to 'be' in order to function in human societies.

The question of language is central to the construction of such 'selves', or subject positions. Some understandings of language are more productive than others, however. Understandings of language as a 'thing' that we 'acquire' are consistent with those psychological notions of personality which need to be problematised. The kind of understanding of language that is more consistent with post-structuralist notions of subject positioning is one which views language as more of a resource than a thing, something that we use rather than have, and something that is always situationally specific rather than general. From such a perspective, there are no general language or communication skills. Rather, individuals, because of the range of contexts in which they have learned to function, may demonstrate a wider or narrower repertoire of specific language or communication skills, learned by effective participation in a wider or narrower range of situations. Such a perspective on language offers women in the workplace a powerful tool, first, for understanding what it is that they have learned to do and, second, for representing their case for recognition of what they do in the most powerful terms.

The perspective on language skills adopted here, and in the What's in a Word project materials, thus involves a considerable widening of current industrial strategies (Frizzell, 1991, for example), in which 'language skills' refers only to skills involving competence in a second language or with written language (reading and writing). This issue is elaborated in the What's in a Word research report, as follows:

> Treating literacy and second language skills as the only language skills, however, and failing to take account of language skills concerned with establishing and maintaining social relations, has two effects. Firstly, it perpetuates a system of industrial recognition of

> skills which gives value to those skills which can be formally accredited and little or no value to those which are learned informally, a system which significantly disadvantages women. Secondly, failing to take account of language skills concerned with establishing and maintaining social relations perpetuates a particular view of the nature of language which has been even more disadvantageous for women.
>
> Language is also widely understood to be a reflection or expression of who people are ('personality') and of what they know ('knowledge'), both of which are regarded as existing quite independently of language itself. From this point of view, learning language is a matter of learning the 'right way' of saying things (with respect to knowledge) or simply 'comes naturally' (with respect to personality). But neither 'personality' nor 'knowledge' can exist independently of language, or some form of communication. Language is not just the clothing we put on something else; language is the means by which knowledge, personality and relations between people are quite literally constructed. We make knowledge, ourselves and our relations with each other through the ways we learn to use language, particularly through talk. (Poynton & Lazenby, 1992, 11–12)

It is hardly surprising then, given the lack of certain kinds of knowledge about language, that the specification of basic competencies in the workplace should treat 'communication' skills, involving spoken and written language, as 'generic' skills, claiming them to be generalisable from one kind of situation to another. While such a view of language skills is widespread, it is erroneous. Language is only a singular 'thing' in the most abstract sense. Actual speakers and writers *use* language (as a resource) in a variety of highly specific contexts and, in order to render it appropriate for that multiplicity of contexts, use it in different and highly specific ways.

Drawing on various aspects of current theories and descriptions of language offers an extraordinarily productive way of going beyond current perspectives on women's 'communication' skills, and of putting more bite into claims for the overdue recognition of those skills. No one should believe, however, that identifying what women do, and spelling that out in words of one syllable, is going to lead to the automatic acceptance of those claims. There is too much evidence that 'skill' is susceptible to infinite redefinition as 'what women don't do', and that employers are less than willing to pay for what they get. A classic example is employer preference for 'experience', which actually means a combination of industry knowledge and sophisticated interpersonal skills.

The workplace interviews conducted as part of the project revealed that, despite their everyday use, 'interpersonal' language skills were not commonly reflected either in the women's position descriptions (where these existed) or in the accounts the women themselves gave of their work. The lack of value given to these skills, together with the generally low status of the positions that women occupy, means that women themselves do not necessarily see them as important as they actually are. The most significant indicator of this is the phenomenon of 'minimising'—women consistently underselling themselves and their skills by saying they 'just' or 'only' did all sorts of things or they did 'a bit' of some highly skilled task.

It is important to understand why women represent their skills and hence themselves in such ways. When women minimise, it is not because the things they do are unimportant but because women themselves are regarded as unimportant. In other words, minimising has little to do with the importance of particular workplace activities and everything to do with the status of those who are responsible for those activities, in the workplace and in society at large.

It is critical that what women say about their work, in the context, for example, of a skills audit, is not simply accepted at face value. The tendency of women to minimise, and for men to 'maximise', needs to be widely recognised. In interpreting women's accounts, it is most important to understand that language does not provide transparent access to an unproblematic 'reality' that simply exists 'out there'. This is of critical importance in providing a response to those who say, with respect to skills audits and interviews: 'But that's what they told us'—as if there is only one way of reading what people say. What people say always has to be understood in terms of who they are, not in the psychological sense of 'personality', but rather in terms of the position they occupy in the society as a whole. Those in subordinate positions have learned to see themselves and the world they live in through the eyes of that subordinate positioning. A large part of how they come to 'see' in that way is through adopting the ways of 'saying' used in that society, ways of saying which come to constitute the way things actually are, rather than being understood as only the dominant ways in which things are interpreted. This is why feminist projects, and those of other subordinate groups, concerned with 'naming' things in different ways, are of such critical political importance: they offer the possibility of people coming to see themselves and their world through eyes other than those of subordination.

Women and men using language: differences, causes and effects

In undertaking the fieldwork for the What's in a Word project, it quickly became apparent that many skills that all too often have been regarded as personal attributes of women—skills such as patience, consideration, friendliness, supportiveness—could far more productively be regarded as very specific kinds of verbal and non-verbal communication skills. The case for understanding such skills in this way begins with the fact that women and men, on the whole, learn very different styles of communication. These styles are not 'natural', developed simply because women are female and men male, but rather they derive from the different modes of socialisation which women and men experience, serving the purpose of positioning then differently: basically, subordinating women to men.

There is now a wealth of research, reported in publications for a general audience (such as Spender 1980, Tannen 1980) as well as a more specialist one (such as Maltz & Borker 1982, Poynton 1985, Thorne, Karamarae & Henley 1983), exploring the nature of the differences between women's and men's ways of engaging in verbal interaction. Women tend to be more oriented to the effective functioning of the group, giving everyone a chance to speak, while men tend to be more oriented towards impressing the group, being more individualistic and holding the floor if necessary.

This fundamental difference in orientation leads to very different ways of interacting. Women are more inclined to make space for others by such strategies as:

— asking questions, thereby inviting others to talk

— not interrupting other speakers but letting them complete what they want to say
— indicating ongoing attentiveness to what others are saying by producing regular signals, both non-verbal (such as head nods) and verbal (including 'Uh-huh', 'Mmm', 'Right', 'Yes', 'I see'), during other speakers' turns
— being less categorical in what they say, for example by using 'may' and 'perhaps' and 'I think', thereby making clear that issues are not cut-and-dried, and encouraging others to add their opinions.

Men, on the other hand, are more inclined to claim space for themselves by:
— telling people things (including jokes and generally showing-off) rather than asking questions, thus limiting the involvement of others to mere acknowledgment of what has been said or to responses such as laughter or, alternatively, precipitating disagreement or even argument
— interrupting other speakers or talking over the top of them and not letting them complete what they want to say
— not overtly indicating a great deal of attentiveness to what others are saying
— being categorical in what they say, presenting issues as cut-and-dried and generally claiming expert status.

This summary of differences is in itself too categorical, as it does not allow for the considerable variation that occurs within the speech of both women and men. Nevertheless, it is useful as a summary of broad tendencies. The implications of such a pattern of differences are considerable. To the extent that women function co-operatively and supportively of those with whom they interact on a daily basis, they are performing an invaluable role in the workplace. In many cases, they will be 'mopping up' the damage caused by the more performance-oriented and competitive male interactive style. In other cases, they will be enabling a male colleague or superior to function far more efficiently and less stressfully than he might with a potentially competitive male in the same role. Similarly, in many work contexts, women interact positively and productively with clients or customers.

One major problem for women is that this kind of work is not visible work. It is not like making something from a lump of metal, or having people doing things at your behest or even producing a large error-free document through the use of a computer. A further significant problem is that any gain for women through acknowledgment of such communication skills may also have less beneficial consequences since the kinds of supportive skills that women are expected to have, indeed relied upon to have, have been learned through processes of socialisation into subordination in a profoundly patriarchal society. Hence, valuing such skills in the workplace might contribute further to cementing women into positions of subordination. On the other hand, the argument that many women, including those in quite subordinate positions, already do the kind of interpersonal work that managers are expected to perform might represent the first step in a long-term strategy for gaining recognition and appropriate recompense for women's language skills.

Language skills in the workplace

Interactive (spoken language) skills

Having good 'interpersonal communication' skills involves being able to interact effectively with people in a range of different work relationships, in particular with customers or clients, colleagues, and managers or supervisors. Women are far less likely than men to be in positions of power over other workers, although they not infrequently find themselves in positions where they have to get others (including people outside their own workplace) to do things for them without being in positions of direct authority over them. In each case, specific strategies involving the use of language must be deployed to establish and maintain the most effective working relationship. There is ample evidence that women operate particularly effectively but that the considerable skill which they have learned to use is not recognised. Alternatively, if their actual skill is recognised, its real value is not understood.

Such lack of acknowledgment derives not only from the general invisibility of women's skills but also from the failure to understand much of what happens when people use language. One example, from the many case studies included in the What's in a Word research report, will illustrate this. Women working as bank tellers often find themselves in a double bind in interacting with customers. On the one hand, tellers are under considerable pressure to complete an individual transaction within a specified time limit. On the other hand, these same tellers are very well aware that they are the 'front-line' representatives of organisations which are selling themselves to these same customers on the basis of 'service' and good customer relations. Good customer relations and the timed transaction sit somewhat uneasily together, however, particularly when you work for Bank X in a suburban branch with a large clientele of elderly people. Such customers expect to be able to have 'a bit of a chat' with the person behind the counter, so that the timed minimum transaction may have the effect of sending them off to Bank Y (or even Credit Union Z) in search of the human face they failed to find at Bank X.

To the extent that retail banking is moving from the provision of everyday banking services over the counter, towards greater service provision via cheaper electronic means, the elimination of 'chat' (and the eventual elimination of the suburban branch) are part of the same inexorable process. From a customer point of view, however, the disappearance of 'chat' confirms suspicions that banks are only 'after your money' anyway, reinforcing widespread cynicism and frustration with current banking services and procedures. This hardly sounds like the good customer relations that banks and other financial institutions purport to understand as central to the conduct of their businesses.

Considerations of power are central to understanding workplace interpersonal communication and language is a key means of negotiating specific power relations. Such power relations are not simply *given* in any particular situation but are *created* in specific ways in and through the actual conduct of interaction between people. In the case of bank tellers, the willingness of an employee to listen to a customer becomes a means by which the customer is (however momentarily) positioned as one whose words have value, as one who 'knows' something that others are willing to hear. By the same token, constraints on the ability of an employee to listen to a customer in this way function as the 'voice' of the financial institution claiming for itself all authority and sole right to speak and be listened to.

Who claims the right to be the sole voice of authority, and how others deal with such claims, are critical issues in negotiating the whole gamut of workplace social relations: both within one's own organisation (between equals, subordinates or superiors), and with 'outsiders' (customers/clients as well as workers in other workplaces). In many situations, it is more likely to be men than women who claim the voice of authority, to be the ones who 'know', whether or not they actually do occupy positions of authority, and whether or not they actually do know. This can create problems, problems which in many cases are minimised by women using specialised interactive skills which enhance co-operation, minimise conflict, mediate between potentially conflictual parties, soften the cutting edge of an abrasive superior's directives to subordinates and suggest more effective ways of doing things without antagonising one who doesn't really know. Such interactive techniques are learned by women as survival strategies in a world in which women are on the whole regarded as less than men. They are techniques that, ideally, everyone needs to be able to use in the service of harmonious social relations and efficient and productive workplaces. The reality, however, is that women are more likely than men to be able to make effective use of such techniques and this ought to be recognised.

Literacy (written language) skills

Just as the existence, not to mention the effectiveness, of women's interactive (spoken language) skills tends to be largely invisible in the workplace, so too are women's literacy (written language) skills. And for similar reasons: first, because these skills are deployed by women and second, because there is insufficient understanding of what is actually involved in the various kinds of writing that women do in the workplace. As with interactive skills, women are commonly in possession of the most sophisticated versions of literacy skills. Girls, as a group, develop greater competence with relevant literacy skills at school, and women are relied upon to possess such skills in the workplace (White, 1986). Again, literacy skills are essential to the productive workplace but are substantially unacknowledged and unrewarded.

Some of the relevant issues can be illustrated by looking more closely at what is typically referred to in the workplace simply as 'word processing', an activity carried out by (largely women) employees using 'word processors'. Such terms carry the implication that it is the machines that really do the work, 'processing the words', with the women operators performing a function similar to machine operatives in a factory: 'minding' the machine, fixing minor glitches in its operation and calling a supervisor when something really serious goes wrong. Such terms also imply that what is at stake in producing documents using word-processing software is simply manipulation of words. Nothing could be further from the truth, on both counts.

Women engaged in using word-processing programs on computers perform a wide variety of tasks, all of them involving both technical skills (with respect to computer hardware and software) and skills with respect to written language. Some of these are extremely sophisticated, involving extensive knowledge of language at every level from the word to the most complex test. The tendency to invisibility of all language skills takes on a particular inflection, however, in relation to *written* language skills. These skills are recognised (to a degree) and are accorded a significant status; employers and employer bodies, for example, regularly bemoan what they see as inadequate literacy skills among new recruits to the workplace.

What such complaints actually refer to, generally, are matters of conventional spelling and punctuation, together with the use of standard grammar in writing. Adequate command of such features of standard written English can be understood to constitute a set of 'generic skills' as far as written language is concerned, but that is only a minor part of workplace written language skills. Just as interactive (spoken language) skills are always specific, involving knowing how to negotiate particular workplace relations in the context of particular workplaces and their specialised knowledges, so literacy (written language) skills are also specific. What this means is that employees engaged in producing written text on a day-to-day basis may need to have an extensive knowledge of:

— the vocabulary (including spelling) relevant to their workplace. This may involve a considerable number of technical terms, which might never be explained or defined but in which fluency is expected
— the stylistic conventions for a range of document types, including formatting requirements and appropriate standard grammar. In many cases, employees are producing documents from drafts produced by others, drafts which may involve only the germ of an idea (which then has to be appropriately worded) or which may be couched in non-standard English or an inappropriate or incomplete format.
— the range of text types used in a particular workplace, involving considerable differences in the structure of texts and hence the relevant vocabulary and grammatical choices. Such differences arise, at the most general level, from the different purposes which texts serve (for example, informing compared with warning, compared with developing an argument) and involve issues concerned with the ways in which written texts, in 'speaking' to their audiences, construct specific relations of power between the addressee and the institutional 'voice'.
— the relations between the grammatical structures appropriate to spoken English and those appropriate to standard written English, where the origin of a text is spoken rather than written (for example, minute notes or dictations). Spoken and written forms of language use significantly different grammatical structures and vocabularies, so that producing a written text from a spoken origin is not a matter of merely 'writing down' what was said.

Writing, in other words, is not simply 'writing', but consists of many skills. And individuals do not become 'good writers' in some kind of absolute sense but rather learn task-specific writing skills. Because much workplace writing is performed by women, because it is learned on the job rather than in formal training, and because of the historical tendency to think of writing largely in terms of spelling and other physical features, the very real skills involved are rendered invisible. Just as women who are characterised as 'good with people' need to be able to identify what it is they actually do in effective workforce interaction, so women characterised as 'good writers' need to be able to identify what it is *they* actually do. Learning how to do this can draw very productively on various kinds of technical knowledge about language, both spoken and written. Such knowledge will assist women to characterise their language skills in the kind of detail which demonstrates that these are indeed sophisticated technical skills. Such knowledge can also be used to assist women to turn their attention to those language practices, such as 'minimising', which undermine their credibility as competent and productive skilled workers.

Conclusion: interventions

The recognition of interactive (spoken language) and literacy (written language) skills as highly technical skills is of particular consequence for women. Women consistently demonstrate high levels of competence in these areas, effectively but without fanfare. Just as consistently, such skills do not receive the recognition they deserve. A first consideration in explaining this phenomenon is that the skills themselves are not well understood. A second, more intractable, consideration is that the evidence indicates that if a skill is more likely to be used by women, then it is less likely to be recognised as skill. Devising strategies to effect change is consequently likely to be a complex and even dispiriting process. Strategies which assist women to recognise and value for themselves, in the first instance, what it is that they do, are clearly of importance. Likewise, strategies which make it more difficult for employers and employing bodies to continue to fail to acknowledge significant areas of workplace skill are also important.

From both these points of view, the materials produced by the What's in a Word project should make a significant intervention. There is already evidence that they will be effective in doing so: a wallchart and practical guide (A Window on Women's Skills in Administrative and Clerical Work) have been on trial in some workplaces with valuable results. Furthermore, the response to the research (from, for example, those interested in aspects of award restructuring and the development of competency-based training for clerical employees) suggests that the project is having an impact at levels other than the individual workplace. There is a sense, however, in which all this can be seen as merely a process of consciousness-raising, which may raise women's self-esteem and their expectations of recognition and recompense for the contributions they make, only to see such expectations dashed. The bottom line remains that skill is historically best understood not in terms of what people actually do, and the real economic value of their activities, but rather, as Barbara Pocock puts it, as 'a product of workplace struggle'. Skill, she argues 'has as much to do with the relative strength of the opposing forces in production—capital and workers—and their competition to control work and maximise returns or wages, as it has to do with technological conditions of production' (Pocock, 1988, 10).

One should not be too sanguine, particularly where the economic interests of women are concerned, in imagining that interventions of any kind will be immediately effective. The likely industrial scenario of the near future in Australia, with its emphasis on enterprise and even individual contracts between employer and employee, does not offer much joy to anyone, least of all women. What one can hope for is that intervention at the level of the individual workplace and the individual industrial officer will make clear to more and more women the discrepancy between what they do and what they are paid for. The increasing politicisation of women workers may offer the best hope of change.

Notes

[1] The What's in a Word project was an initiative of the Women's Adviser's Unit of the South Australian Department of Labour. It was jointly funded by the South Australian Department of Labour and the Commonwealth Department of Industrial Relations through the Workplace Reform Program. Fieldwork was conducted by Kim Lazenby, Senior Project Officer, Women's Adviser's Unit, and the author. The project has resulted in three publications: *What's in a Word: Recognition of Women's Skills in Workplace Change* (a Report, Wallchart and Practical Guide). Copies of these publications are available from the Women's Adviser's Unit in South Australia and from the Department of Industrial Relations state offices elsewhere. The original seminar in the CIRCIT/URCOT program was presented jointly with Kim Lazenby. My thanks to the Women's Adviser's Unit, and particularly to Kim Lazenby, for their ready agreement that I should publish this paper independently.

6

Human-centred systems, gender and computer supported co-operative work

MIKE HALES

This chapter maps some of the shifting ground in the design of computer systems, and its implications for work organisation and women's skills. I discuss some of the limits of the British and European human-centred systems movement, from a personal background of fifteen years involvement, and with a feminist perspective. I suggest that it is no longer necessary to accept the movement's limitations because some alternatives are emerging in the sphere of technology design and use. In particular, a movement emerging around Computer Supported Co-operative Work (CSCW) makes available different actors, traditions and objects to work with; it offers a different disposition of resources, locales and locations for action. It is not specifically woman-friendly, but CSCW people get closer to asking some of the key questions highlighted by feminism.

The human-centred systems movement (or the 'HCS' movement) has focused mainly on skilled manufacturing work, and thus on male workers. The CSCW movement is not constrained in the same way, which is part of its interest. In addition, the movement has a particular interpretation, a particular politics, of how humans can get from the margins to the centre of things. My perspective derives from yet another politics—of challenging 'universal' and 'centred' with 'local' and 'de-centred' knowledges and practices. It is about deconstructing and radically reconstructing the intersections of things and discourses and actors. This other politics has feminist (though not exclusively feminist) origins.

In this sense, the chapter attempts a gender-aware critique of the human-centred systems movement. This movement has never made any particular claims about dealing with gender as a central concern. Empirically, it mainly addresses male-dominated areas of work and responds to the demands of male-dominated institutions (engineering trade unions, academic work and government), seeking to defend and develop traditional (gendered) forms of job-control and skill. A truthful and necessary critique of HCS might be developed on these grounds—for example, illustrating its neglect of office work, where a majority of technology users are women, and the significant presence of patriarchal forms of technical, academic and labour movement practice.

However, my purpose here is to raise the question of gender in a different way, by focusing on ways of knowing and acting. Feminist thinking has developed a whole range of issues about knowledges and their locations, knowledge and action, personal identity in relation to 'universal' forms of knowing and acting, and the politics of the very notion of 'centredness' in knowledge. There is a critique, in other words, of the genderedness of certain ways of knowing and acting. My reason for being interested in CSCW is that it contains some of these 'theory and practice' threads, in ways that the human-centred systems tradition does not. However, these threads did not always get into CSCW by feminist routes, so some care is needed in reading the gender-political significance of CSCW practices and discourses. Basically, my claim is that CSCW constitutes important and—compared with HCS—possibly better ground for feminists to work on; not that CSCW's practices or CSCW technologies and systems are intrinsically woman-friendly. The claim can be no stronger because, of course, no technology and no method can resolve the problems of struggle, on real ground.

Things, languages and powers

This analysis is driven by a single, difficult question: Why don't we do more things different? Living and working in worlds where so much is the wrong shape (for most men and for most women), how is it that we reproduce so much of the misfit, at such personal cost, investing in our own continued suffering? Why do we run in ruts that sometimes are taller than we are? At one level I interpret this 'doing things different' question as being about the apparatus of our working lives—technology. In a strong sense, apparatus (systems of technological artifacts) can be seen as being the ground that we move in, in a literally constructed society, and in this respect the 'things' that need doing different are actual things, physical artifacts. This is the ground on which the human-centred systems movement pitches its struggle.

But at another level my 'doing things different' question is also about cultures and about what Bourdieu (1980) calls 'habitus', a kind of culture-as-body-language concept. In other words it is also the doing that needs to be different, the ways in which making and using things is enacted. Thus the 'doing things different' question is about the subjective as well as the objective forces which maintain practices in pretty much the same shape from year to year, generation to generation. It is a question of unwitting or unwilling conservatism. It is about being able to speak of the not-obvious in the obvious, so as to be able to construct something different which differs in all the necessary respects. And so, while discussing different 'alternativist' territories available to work on within technology design (human-centred systems, CSCW), this chapter is also about a choice of languages for speaking of different landscapes and different futures. The basic argument is that a post-modern–feminist language is more powerful than a 'human- centred' systems-centred language, and that there are practical connections for the former with CSCW which ought to be explored and developed.

The combination of objective and subjective is crucial in dealing with the 'obviousness' of technology. The underlying link between the two is found in hegemony. This term invokes a concept of how power is produced and reproduced without force being apparent, and is particularly helpful in dealing with the politics of technologies, where power is mediated through things and non-human processes. Human-centred systems thinking takes the possibility of counter-hegemonic practice for granted, seeing it as an ordinary aspect of the skill, ingenuity and creativity of ordinary people. Post-modern feminism is rather more circumspect, and concerns itself with the conditions under which the ordinary can actually be a basis for political practices and alternative visions can actually become ordinary—that is, hegemonic.

From here this chapter proceeds as follows. First there is an outline of how the HCS ground and the CSCW ground relate to each other. Then there is a critical, participant account of human-centred systems practice in the Greater London Council in the mid-1980s and a shorter interpretation of current 'Euro' HCS activity followed by a commentary on difficulties inherent in the human-centred systems vision of 'human'. Finally there is a brief survey of the approaches of some feminists in CSCW and a discussion of some issues in evaluating CSCW in relation to the human-centred systems movement, as alternative spheres of socialist–feminist activism.

Some history

The idea of 'human centred' alternatives in technology, twinned with 'socially useful' alternatives in economics, emerged during the 1970s in British trade union struggles in an aerospace multinational (Wainright & Elliott 1982). The original actors in human-centred systems design were male engineering craft-workers and professionals, activists in engineering trade unions. During the late 1970s in England the strategy of the Lucas Aerospace shop stewards' combine committee was very much the Good Thing that was happening somewhere else, tracked with great interest in academia as well as by trade unionists. Eventually the combine lost its fight, leading members were sacked, and in 1982 the twin ideas of human-centred systems and socially useful production came to rest in a new location, the newly formed Industry and Employment Branch in the Greater London Council (GLC).

This was quite different from the originating context of struggle within a hi-tech manufacturing multinational and its formal trade union structure. The vision of socially useful production had originally been articulated from the 'worker' side (non-management side) of manufacturing. With the GLC initiative the terminology was suddenly transplanted to the 'management' (policy-making) side of the economy, in the local state (Cockburn 1978). From 1986, when the GLC was abolished by Act of Parliament, human-centred systems practice began to move into the Euro-funding sphere, with ESPRIT and other European programme funding of projects. Now, in the 1990s, the European HCS movement has substantial German, Scandinavian, British, as well as Japanese strands, and has become closely associated with strategic debates about post-Fordist or neo-Fordist industrial competition ('flexibility', 'quality') in manufacturing (Cooley 1989).

Over the same period, emerging first in the US, there has appeared a movement concerned with the design of software aimed at group rather than individual use, with group processes and with 'user-centred' and use-centred designs and design processes. One major strand is the design of 'friendly' interfaces, following the success of the Apple Macintosh concept. Another is participative design. A third is the experience of computing professionals in communicating through e-mail and other messaging and networking technologies, leading to the notion of 'groupware' as a distinct category of software applications.

The actors in CSCW are information systems and computer science professionals, and others from social science and cognitive science disciplines, in academia and in industry (in for example, the corporate labs of Xerox and Hewlett Packard). CSCW is a genuine social movement (for example, it mounts a challenge to established ways of doing things in the economically significant sphere of information systems development), but in a micro rather than macro sense, a movement of technical professionals within their own domain rather than on any broader social or political terrain.

Although the term CSCW originates in the US, the CSCW community overlaps with the European human-centred systems community with respect to Scandinavian traditions which are common to both; the 'tool based' approach to participative, trade-union-oriented design of computer systems, (for example, Ehn 1988). Nevertheless the main concerns and actors in CSCW and Euro HCS are different. In particular, CSCW has never been an explicitly political movement, and CSCW centres on office work (managerial, administrative and especially technical–professional) rather than factory work.

Given its different focus and membership, CSCW has produced insights into design that fall outside the human-centred systems tradition. One analysis (Hales, 1992) suggests that three styles of information systems development may be seen emerging in CSCW. Characterised in terms of differing relationships with 'users', these are: users as clients, users as co-designers and users as illegitimate actors and constructors. On this analysis, the significant new ground of CSCW lies mainly in the third dimension, and CSCW's main overlap with the human-centred tradition lies in the second ('co-designer') strand.

A couple of years ago, it had begun to seem to me that participative design had begun a steady slide from the self-consciously political ground of trade union oriented 'human-centred' design into the mainstream professional repertoire of design technique. There was, for example, a regular and increasing confusion of user-centred and human-centred. Nowadays participation can mean little more than 'thorough knowledge extraction', enabling richer specifications for systems with higher degrees of 'usability'. It was especially worrying that such a shift seemed quite consistent with the product-focused emphasis of the human-centred systems tradition. In one situation that I knew, a serious attempt to develop a major database system in a woman-friendly and human-centred way, with 'participation' as the keynote in design, had probably yielded a very good system product but no clear-cut guarantee of greater benefits for women in use (Hales & O'Hara 1993). So I had begun to look more closely and critically at the 'co-design' ground that CSCW shares with the HCS movement.

The third 'actor/constructor' strand of CSCW supplies a kind of voice which HCS does not seem to have, and through this, a possibility of addressing issues that 'participation' and co-design can ignore. The idea of users of technological artifacts as 'actors', agents who 'construct' the practical realities in which they act, derives from a broad tradition of subjectivity-oriented theory. This tradition has a common commitment to approaching 'the obvious' as both complex and constructed rather than simple and 'natural'. Within this tradition it is not possible to take 'human-centred' as an attribute of things; there could, for example, be no human-centred machine or piece of software. Human-centredness would have to be seen as inhering in action; it would be necessary to talk of human-centred practices rather than human-centred systems. Given the possibility that product-centredness in 'human-centred' systems practice may be blocking further development of its political aims, this action-centred emphasis in CSCW is very welcome.

Things and strategies in the human-centred systems tradition

In the GLC adventure, human-centred thinking was dropped into a relative political and economic vacuum. Not surprisingly, it exploded. In Lucas Aerospace a discourse on human-centred tools and socially useful production values was firmly located within the political and practical framework of collective bargaining in a given hi-tech manufacturing multinational. This gave it a certain concrete legitimacy and practical/political content. But 'London' has none of that local coherence—political, practical, technological, economic. London is not a locality at all, but a complex and shifting conjunction of very many locales; economic, cultural, ethnic, political; at global, national, regional, city, town, village, family and work group levels. It's hard to see how there could be a strategy of any kind for 'London', let alone a technology strategy.

In this different and difficult setting the GLC entrusted its financial investment arm, the Greater London Enterprise Board (GLEB), with the development of human-centred systems. Mike Cooley, an ex-Lucas trade-union activist, was GLEB's technology director, and the agencies formed to do the work were called Technology Networks (TechNets).

Although socially useful and human-centred production served as rallying cries and were taken to express some central values, this language was too weak to clearly structure the directions in which TechNets would go. The pair of slogans was unstable; each had a double connotation. Thus, depending on who was speaking, we had either socially useful products or socially useful production; and either human-centred systems or human-centred designing. The 'human-centred' language was used by different speakers to focus attention either on things—designed, engineered, manufactured artifacts—or on the practices through which economic goods (both things and services) come into existence.

Previously, in the Lucas campaign, the language had tended strongly towards 'things': alternative vehicles which could use energy efficiently or escape the constraints of specialised first world transport infrastructures (roads, rail networks); alternative advanced tools which called on traditional skills; alternative living equipment for disabled people. This was pivotal to the Lucas Plan style: show that alternatives are technically feasible, and thence prompt awkward questions about political and economic feasibility. It was called 'technological agitprop'.

The characteristic product-centred practice of the human-centred systems movement pivots on tactile, tacit, tool-focused experience. While humans certainly do act on the world in a tool-centred way, they also act in the subjective world—and in this way they define (constitute, produce) the objective world—in language and in culture. It seems an especially serious problem if a progressive movement like the human-centred systems movement chooses (in contrast with the localism movement, for example) to address tools and the production of things to the exclusion of texts, languages and the production of cultures, lifeworlds and identities.

In the TechNets this product-focused tendency was reinforced by a stereotypical division of labour emerging within GLC/GLEB: non-engineers in the GLC were often seen from GLEB's technology division as 'critics' and 'sociologists'—Sociologist Bad, Engineer Good. Technologists were recruited in GLEB to oversee TechNet strategy, while voluntary sector activists, feminists, politicos and innovation researchers—'sociologists' in general (who were inclined to take a socio-economic, cultural, group-developmental or agitational view)—were sequestered at the GLC. TechNets were expected to produce not practices ('networks' or communities) and certainly not analyses or interpretations, but literal things that could be photographed and campaigned about. Apparently, these were 'human-centred' things; there was an expectation that somehow, determinate social and political relations could be coded into, or triggered from, artifacts.

The interpretation of social usefulness in TechNets was tied to the idea of 'wasted resources'. In the early 1980s there was a widespread commitment to resource centres of various kinds, support for local organising in opposition to the centralist national state and transnational industry: law centres, trade-union resource centres and so on. Dutch streetfront 'science shops' linked with universities were a prominent model. Thus, during the Lucas Aerospace struggle the trade union Combine had sought access to academic resources (staff time, equipment) in order to develop prototypes for technological agit-prop. This history was generalised in the TechNets, which were defined by GLC strategists as institutions which would key into higher education institutions across London, liberating equipment and staff time which were 'under-used' and kept from the public by the protocols of academic practice. By institutionalising this academic connection, TechNets took a step in the professionalisation and 'professor-isation' of the human-centred systems idea. Academics and professionals rather than 'ordinary people' became the de facto core of the resource-led networks set up by the GLEB.

Even at the time it was possible to see problems in this. For the Left in the 1960s and 1970s, central terms of debate had included the 'new working class' of technical workers, and the 'embourgeoisement' of the traditional working class—both of them confusing themes, awkwardly stuck in the framework of economistic Marxism. By the early 1980s, alternative approaches to class were available, such as Illich's critique of professionals' monopoly power (Illich, 1975) and the socialist–feminist debate in the United States about 'the professional–managerial class' (Ehrenreich & Ehrenreich 1977; Ehrenreich & English 1979; Walker 1979). Such approaches offered critiques of the contradictions underlying 'radical professional' movements of the 1970s.

Thus it was problematic that 'ordinary people' should be adopted as the TechNets' central political term. It apparently embraced professionals and academics—or some of them. It was not quite clear who was 'ordinary' and who was not; real humans, it seems, would know them when they worked with them. Essentially the term seems to mean people who are not bosses or don't act like bosses. Such a Popular Front concept might almost have been designed to obscure the reality which is hard to name, but which can be called the professional–managerial class, and to legitimate a leading role for academic professionals. In the absence of 'demand-pull' grassroots linkages between locales in 'London', connections often got made through a kind of 'supply-push'. We radical professionals at the GLC and in the TechNets, with our local–state financial resources and 'liberated' academic time found ourselves bringing the human-centred idea and the unasked-for technological agitprop resource base to ordinary people—if they could be enticed into our meetings and our buildings. Given that 'network' was the organisational metaphor, this was an embarrassingly 'centred' process.

Since the TechNet experiments of the mid-1980s it has become clear that there are many national and local HCS cultures (Gill 1990; Badham 1991) and that these vary in contents and directions. However, a sort of Esperanto has evolved in Euro-funded, multi-country initiatives, a vocabulary of 'skill-based', 'human-centred' and 'anthropocentric' systems. The terms are widely interchangeable (although the German term 'anthropocentric' is the central one, reflecting the strength of German institutions at the Euro-political level). All three terms refer to 'systems' rather than to design or development practices; in the 1990s the things/practices ambivalence of the earlier movement seems firmly settled in the 'thing' domain.

In human-centred systems thinking, culture (and the problem of acting with people) is pushed to the margins by systems (and the problem of acting with things). Twenty years ago, as the movement was emerging, other groupings—feminists, the New Left, the libertarian Left—were struggling to deal with problems of politically organising with 'ordinary people'. But these other traditions addressed culture and identity as central, refractory political realities, structures of external and internal meaning which must be addressed by political activists, and they gave them serious intellectual attention. The human-centred systems tradition (the systems-centred humanist tradition?) barely has a language for talking about culture. Where it goes into any detail at all, it tends to founder on problems generated by the tradition's own professional-centredness: at the level of interdisciplinarity, crossing the borders between engineering and social-scientific academic cultures.

What is 'human'?

There are a number of important things about human action in the late twentieth century which the human-centred systems tradition probably cannot express. The deepest difficulty is the movement's politics of obviousness. This derives from a political identification with 'ordinary people' but tends to a refusal of difficult language, language which goes beyond the obvious. Such a commitment makes theory of any kind problematic and obscures the task of radically deconstructing our present so that we may make a future that is different in all the necessary ways. This is the kind of shift that feminist thinking in recent years has made, from an empirical focus on women-and-men to a focus on the genderedness and gendering of things and roles. Of course, culture and identity are hard to understand in a theorised way; but this is not really an acceptable excuse for focusing on tools and presuming, as human-centred systems thinking does, an obvious framework of meaning in tool-using and tool-making situations.

The 'human-centred' commitment to obviousness and tacitness is confounded with an identification of Taylorism as the political enemy. Taylorism is a tradition of design which systematically detaches 'scientific' knowledge of work from those who actually do that work. Having identified the enemy in this way, the human-centred systems tradition runs the converse risk, of reducing humans to 'hands' by refusing to recognise the legitimacy of 'head' knowledge (knowledge which does not live entirely within immediate hands-on experience, but moves 'out there' in abstracted and often disembodied forms—for example, libraries or, generically, texts).

The human-centred systems tradition is stuck with a confusion between a historical class practice, Scientific Management/Taylorism, and a basic human necessity—putting knowledge 'out there' in one objectified form or another, in order to communicate at all. Turning its face against the former practice, the human-centred systems tradition has failed to think hard enough about the political implications of the latter, general condition of human action and emancipation. A politics of obviousness, which has 'ordinary' people working with other 'ordinary' people, presumes too much about what people are able to leave behind when they move into a new situation. Because of the role of obviousness in the marginalisation of women ('Obviously women cannot do x . . .'), feminism is conscious that obviousness is not a safe base for building alternative practice—even when it is a necessary base.

A second difficulty comes with the central methodological notion of the human-centred systems movement: participation. Participation implies access to existing Objects within some existing structure, and the metaphor of 'centredness' then implies moving an existing Subject from a periphery to a centre, in a landscape whose geometry stays the same. As a working definition of human-centred practice, 'participation' invokes a hermeneutic process—a process of mutual understanding. However, the 'access' idea behind participation is not enough to mobilise the kind of politicised and theoretically developed awareness of systematically distorted communication that belongs, for example, to European traditions in critical theory and post-structuralist theory. The latter recognise that Subjects and texts (readings, meanings) are deeply structured and historically constructed. They can not be simply taken over into new situations; nor can they necessarily be changed overnight (through participation in some new group practice, for example). The human-centred systems tradition simply seeks different, existing humans—non-bosses rather than bosses—at the centre. But how can it deal with the fact that bosses are humans too, have cultures, are skilled, have needs? The rhetoric of participation and social usefulness is simply not political enough. It's unable to speak (in the way that Critical Theory, or Gramscian hegemony theory, or feminist theories of decentred subjectivity attempt to speak) of how humans may legitimately and effectively displace other humans from the centre.

In other words, location is a problem that is absent for the human-centred systems tradition. How humans may legitimately displace other humans; how one practice's knowledge may be legitimated in another—such problems vanish. Particularly, the tradition neglects problems of mapping and taking up positions on the politicised ground of professionalised work. Who are 'ordinary people'? Are men 'ordinary' in the same ways as women? Are engineers and academics ordinary people when they sit in a room with non-professionals and non-bosses? Are the system-centred values and lifeworlds of engineers in contradiction with the practices of some other people; or does the status of all non-bosses as 'ordinary people' somehow resolve the problem of conflict and contradiction? Traditional Marxism deals with such questions through the discourse on class location, which typically sees class as an objective relation. Socialist–feminist theory recognises that not only are men and women not identical, but that neither are women. This recognition has both extended and challenged the Marxist tradition by exploring subjective relations between heterogeneous practices and actors. In contrast, the human-centred systems movement generally adopts voluntarism—willingness to work participatively on system design as an 'ordinary' person—as a substitute for any kind of political analysis of social and historical location.

Materialism is another area where the human-centred systems movement has problems. Its materialism is typically quite vulgar—a focus on action through 'systems' (a sophisticated way of saying 'machines'). This has political and cultural origins: on one hand, a political reaction to Taylorism's and capitalism's transformation of work through redesigned and redistributed instruments and objects of labour; on the other, the taken-for-granted male lifeworlds of skilled craft and professionalised engineering work and trade union activism in manufacturing industries. If 'sociologists' constituted one pole of a distinction in the 1980s GLC, those at the alternate pole might reasonably have been labelled 'tool-makers'; this has exactly the right (labour aristocracy) resonance in British labour movement culture. The history and future of tools is complicated, as the Scandinavian 'tool-centred' tradition recognises: for example, Ehn(1988) draws on Heidegger, Marx and Wittgenstein. But even such sophisticated variants in human-centred thinking marginalise the recognition of texts as distinct from tools in human history.

An adequate materialist approach to work—including engineering work and especially work with computers—needs to be able to address both tools *and* texts. This has an obvious significance for women, many of whom work with technologies that are defined as text-processing. But it goes deeper: every artifact is also a text, carries meaning; some tools (programmed tools) are constructed literally as texts (software); and some tools present themselves through textual interfaces. Approaches which either focus on text (such as hermeneutics or post-structuralist/post-modernist analysis) or on the relations between the physical/corporeal and the symbolic (such as the anthropological materialism of Bourdieu) open up ground for human struggle which is obscured by an excessive focus on tools. For socialist–feminists the central issue is not that women work with text-processing systems, but that socialist–feminist theory has tried much harder to handle the full complexity of the material world—tool and text—and thus is closer to being able to deal with the specific characteristics of information technologies.

A final constraint on human-centred systems practice, closely linked with the politics of obviousness, is traditionalism. Much of the emotional and thus practical–political force of the movement arises from a recognition that valued areas of tradition—notably associated with craft work and craft products—are being systematically devalued and destroyed. The focus is on both the perceived and perhaps irreversible loss, and on the fact of rapid disorienting change as a danger in its own right. Anger rooted in this way is an important political resource. Human-centred systems thinking, however, tends towards a yearning for stable culture, and falls easily into a fondness for a 'golden age'.

To the extent that human-centred systems thinking does recognise culture, it typically reduces culture to tradition (as in the deference reserved for ethnic and regional 'cultural diversity'). Tradition is 'naturalised' culture; but culture is, precisely, the sphere of the unnatural. Culture is where we constitute ourselves as different from nature; the risk of losing both ourselves and nature is intrinsic in the human condition. Culture is where we invent ourselves and our worlds, and is a process which—like subjectivity—is never completely available to any actor. This is where the radical 'cyborg' post-modern socialist–feminism of Donna Haraway, for example (Haraway 1985), is so much more courageous than the comfortable naturalism of human-centred systems artificers.

The human-centred systems tradition presumes that we are human, that we know who, where and what we are, and that the problem is to be in a different place, rather than to be different. And thus the struggle to become human is lost, in a simplified 'humans from the margin to the centre' struggle against Taylorism and bosses. Feminism, of course, cannot be complacent about 'knowing who we are', and thus thinks much harder about subjectivity seen as a process which is larger than we are; about ourselves as something less than we otherwise might be; about 'being human' as becoming more human. On one hand, this gives feminism a much larger political world to act in. But on the other, it makes politics much harder work. A 'systems' rather than a 'systems-and-lifeworlds' view of politics is definitely a softer option.

Perspectives on computer supported co-operative work

Although they have overlaps, the CSCW movement differs in significant ways from the human-centred systems movement. The former is about information systems design rather than manufacturing systems, and thus must address professional and white-collar work as a central concern, including much work which is stereotypically female, from librarians to inquiry desk staff to personal assistants to data entry clerks. CSCW is about work rather than mere systems, and this concreteness requires it to address people in situations including, perhaps, whether the people are women. As an intellectual movement CSCW draws on traditions of thought and practice outside the human-centred systems tradition—notably ethnomethodology—which enable designers to seriously address problems of 'text', and combinations of tools and texts within practices, and relations of actors with text-artifacts. Overall, as a debate about systems design CSCW is in a certain sense more critically committed than human-centred systems thinking, less willing to settle for common sense even while recognising the practical significance of 'obvious' interpretations and relations.

Of course CSCW is not a panacea; it has its own contradictions and eventual shortcomings. For example, the 'thing' fetish is active in CSCW; how could it be otherwise in a sphere of advanced technology? Many members of the CSCW community (for whom 'computer supported' seems to be the primary term) see CSCW as a kind of stuff: a technology, a generic product domain of interfaces and distributed computing systems, a challenging specialist sphere in which the Taylorist mission of software engineering is to be pursued with cognitive science insights, a marketing opportunity for shrink-wrapped generic 'solutions'. But there are others (who primarily address the 'work' term, and the even more problematic 'co-operative') who see CSCW as an approach to production rather than simply products: how are knowledges produced in design and use practices? How are artifacts produced by knowledges? How are practices produced around system artifacts? These are some of the questions—methodological rather than technological—that CSCW researchers are now asking.

Unlike the human-centred systems movement, CSCW has a number of prominent feminists and theoretically articulated feminist arguments. Here I will briefly sketch some of the differing approaches that these scholars are developing within CSCW, all of which contrast in some way with the dominant style in the human-centred systems tradition.

The most direct CSCW feminist connection with the human-centred systems community is through women involved in Scandinavian participative design traditions. Joan Greenbaum, for example, has approached systems design as a practitioner and as a socialist–feminist organiser in office work, via (United States Braverman-style) labour process critiques and via gender analysis drawing on Evelyn Fox Keller (Greenbaum 1990). An American, she works with Scandinavian colleagues developing approaches to co-operative design within the European 'collective resource' tradition (Greenbaum & Kyng 1991) drawing on the UTOPIA project (Ehn 1988) and other Scandinavian tool-oriented practice. She is now exploring what these traditions might mean in the different cultural and political context of the US (Greenbaum forthcoming). The US lacks, for example, the industrial training, industrial democracy and social-democratic traditions of Denmark or Sweden. But Greenbaum's work differs from the European HCS in at least two ways. First, she approaches with the interest of an organiser—a reconstructer—rather than a conservator. And second, as a feminist she is sensitive to the subjective—for example, storytelling as a design method and a form of political practice. Other women from the Scandinavian 'computers and democracy' tradition are in CSCW too, such as Susanne Bødker (see below), and Gro Bjerknes and Tone Bratteteig who worked with nurses on systems design in health care (Bjerknes & Bratteteig 1987).

This is perhaps the most obvious kind of feminist approach in CSCW, drawing directly on well-established 'women's values/ women's needs' interpretations of feminism. For me, these CSCW-feminist approaches are the most familiar because they have been the main reference for the Human Centred Office Systems Design project at Sheffield Polytechnic (Green, Owen & Pain 1991, 1993). Action research focused on office work has been marginal on the British map of human-centred design; in particular, the feminist-inspired Scandinavian practice of study circles (Vehvilainen 1986), as explored by the Sheffield group, lies outside the mainstream of British HCS work, which operates more in terms of multi-disciplinary expert design teams. In contrast CSCW, with its office-work focus, is beginning to look at ways in which 'non-co-operative' commodity tools such as word processors and spreadsheets can be appropriated as a basis for office workers' co-operation (Clement 1991; Nardi & Miller 1991; Gantt & Nardi 1992).

I was much less prepared for another strand of CSCW thinking, which explores the notion of action in relation to technological artefacts. A widely cited formulation of this interest in action is Lucy Suchman's distinction between 'plans' and 'situated actions' (Suchman 1987). Originally her critique was located in the artificial intelligence domain, where rule-based action is widely presumed to be the norm; she challenges this with the anthropological insight that the most general condition of human action is 'situated' recognition based on a low level of reflective self-awareness, rationality in the rule-based sense being the (significant) exception.

Suchman's critique of plan mentality is directed at interface design (user-friendly photocopiers, actually; Suchman is a researcher at Xerox PARC, Palo Alto). But it can be read as applying to the whole rationalist project of design, typified by the present strong professional drive towards 'engineering' status for information system designers, and linked with increasingly formal, certifiable, top-down, sequential, programmatic knowledges (such as 'structured' development methods, or 'formal methods').

In the CSCW literature, Suchman does not seem to make the feminist, as distinct from the anthropological, connection with 'situated' action. But in her closing remarks at a conference on women and computing she noted that:

> Characterisations of computing as inherently masculine or feminine, hard or soft, logical or intuitive are giving way to detailed analyses of the relation of specific computer technologies to particular ideologies and experiences. Similarly, characterisations of gender are opening out, from lists of opposing traits assigned to a priori categories of male and female, to complex analyses of forms of social identification and exclusion and their consequences (Suchman1991, p. 431).

Like Joan Greenbaum, Susanne Bødker is involved in co-operative design (for example, Bødker & Greenbaum 1988). But her work has a broader materialist focus than the tool reductionism of the human-centred systems movement, drawing on 'activity theory' (Kuutti 1991b). Activity theory is concerned with relationships between 'tool' action and praxis/cultural action. This dual tool/language interest is especially important in dealing with computers because, more than other types of machine, they exist in both the physical and the symbolic domain; they can drive a mechanical machine such as a welding robot or an unstable aeroplane, and they can also structure literal dialogues with and between people. Activity theory offers concepts and schemes for thinking such tool/language relations, and information systems researchers have proposed the 'activity' concept as a basis for developing design approaches in CSCW.

Bødker (1991) uses activity theory to generate a theory of practice, and in this respect seeks to go beyond a naturalistic focus on tools. At present this kind of approach does not seem to have any specifically feminist focus. In its socialist–feminist connections through researchers like Bødker, it at least remains open to issues of design work with women, and broader issues of power.

A concern with agency and structure appears in the approaches of two further women in CSCW, Marja Vehvilainen and Susan Leigh Star. Vehvilainen (1991) criticises both 'objectivist' perspectives (which take structure as static, given: patriarchy, women's values, organisational forms and objectives, data structures) and 'subjectivist' perspectives (which can see only individuals and their individual actions—women/men, these users/those users, these designers/those designers). The human-centred systems tradition oscillates awkwardly between the objectivism of designing 'alternative' things and the subjectivism of 'the ingenuity, creativity, etc of ordinary people'. Vehvilainen suggests that structuration theory (Lyytinen 1990; Lyytinen & Ngwenyama 1992) is a promising perspective in social theory, which attempts to address the duality of action, both determined and determining, alive and embodied in artifacts. Vehvilainen's starting point is that 'in order to perceive information systems as gendered, one first has to see that information systems are social'. Thus she seeks a way of thinking about artifacts/structures as mediators of action/actors which also is able to address gender as a mode of sociality.

Susan Leigh Star is concerned with agency and structure as embodied in knowledge artifacts such as artificial intelligences and expert systems. In a discussion of 'invisible work' (Star 1991) which links this feminist concept with the interactionist sociological concept of 'articulation work', Star weaves a number of feminist themes: 'articulation work' itself (the active and material but implicit constructing of a social whole by actors); the 'insistence' of standards which cause small and large sufferings, as a basis for a critique of universalised knowledges and rules; the painful and powerful vision of outsiders. She addresses work as 'the link between the visible and the invisible', which is, I think, another way of putting Hannah Arendt's insight that through work, humans become immortal.

Such approaches offer perspectives on work as the intentional modification of our lifeworld(s), the embodying of subjectivities, the constructing of intentions in and among networks of artifact–agents. On the one hand such perspectives support my long-standing expectation (Hales 1980) that some kind of critical labour process theory is useful for approaching practical problems of power (including patriarchy), identity (including gender identity) and technology. On the other hand, the whole point of such theorising is to enable us to read the authorship in the technology text and also—or why bother with theoretical work at all?—to write and rewrite those texts. The rigorous requirement of this kind of theory (theory-of-practice) is that it must be available in the practice of reading and writing technology, at the point of action.

This is difficult, but it is a difficulty that feminist theorists recognise and struggle with. Star, for example, comments: 'I developed a habit of trying to restore the agency to whatever I read'. This determination to refuse the obvious, and to recognise historical practices and agents in 'nature' and in artifacts, is what I value most in feminist theory and in the more radical strands of CSCW.

Does CSCW have a politics?

Of course it does. If feminism has meant anything over the past twenty years, it is that every practice is a politics. The question is, does CSCW have a significant politics, does it offer significant new ground and new identities for struggles, especially by women?

In the first place, CSCW is significant because it is intrinsically about global economic restructuring; it is an emergent part of the technique of making global restructuring practical. CSCW is about time-space and the division of labour (who does what, where, and when, across the face of the globe), about big organisations and knowledge workers (who, respectively, will pay for and work with CSCW systems), about the strategic investing and configuring of technological capital, about new varieties of generic software ('groupware') and the commercial struggle to be first in future markets with world-sweeping products of spreadsheet-style generality. CSCW is in there where some of the main action is, as future generations of technology are laid down as a substrate for growing future generations of work. CSCW is about the future, in a way that the human-centred systems vision (especially given its pre-modern romantic tendencies) is not. Thus, it is an important site for critical analysis and for the promotion of alternative practices.

CSCW is also important because it promotes a much stronger focus on local configurations of technology, as distinct from generic ('universal') systems. For example, at one extreme, there is increasing talk about a future generation of applications software being 'user enhanceable', which means that software producers are looking towards new types of generic software designed to be configured locally to meet local requirements, with a minimum of local technical knowledge called for. At the other extreme, CSCW researchers have begun to examine the ways in which end-users of current software, not designed to be used co-operatively, do in fact configure local usages in a co-operative way.

In between the extremes a great deal of professional attention is being paid to design for 'usability', to participative or co-operative design with users, and to understanding actual group processes in situations using information technology. All of these trends place emphasis on the kinds of design activity that can appropriately take place in 'use' situations; gradually, an understanding is emerging—from the professionals' point of view—of the practicalities of regarding users as designers. This does not mean that all users of computer systems will in future automatically be offered a role as active designers. But it does mean that new ground is emerging on which women can make demands as users of computers. The more that design professionals become aware of the difference between generic and local design (including both design in use and design of use) the less legitimate it will be for systems to be dropped on users in the time-honoured, Taylorist style: 'Here, work with this now, we know best.'

The third 'plus' is about feminist politics in the professionalised design community. The traditions that are active in CSCW include some which merge with the established human-centred systems community and some which link self-consciously with feminist theory and activism. It is good news to find that some of the active debate among design professionals at an international level, about the design of advanced new technologies, is self-consciously drawing on political traditions. Within the CSCW debate it is possible to speak of a range of difficult and significant things which escape the human-centred systems tradition's anti-theory, pro-craft work and pro-engineering bias: 'location', the localness of knowledges and the practical problems of being present in more than one location; the power of outsider knowledge and the challenge of establishing forms of de-centred power; practical relationships between theoretical rigour and everyday practice (deconstructing the obvious but staying close to action); texts as well as tools as mediators of human development. The feminisms and other radical intellectual approaches which I am invoking here are, quite simply, braver than HCS. More prosaically, in CSCW it's 'natural' to talk of office work as well as factory work.

However—this is the first 'minus'—CSCW has a strong component of California dreaming. Many of the members of the CSCW community are so hooked on the technology, and so wound up with the privileged and technologically well-resourced conditions of their own professional work in the hi-tech sector, that they cannot see the global 'work' wood for the Silicon Valley trees. A large part of the CSCW community is not especially interested in the debate about design, its objects, the social, political and gender location of designers or the legitimacy of all of these; they simply want to work on the latest, seductive technology (multimedia, for example). There seems to be no developed geopolitical or economic angle to CSCW's internal debate; it is still only a debate about style, as distinct from power, location or leverage.

The concern with style is significant. For example, it allows questions to be asked about the relationship between users and designers (where are they, and are they women?), the real needs of different real users (generic users in areas of stereotypical 'women's work', for example), and the actual nature of 'co-operation' in actual work (in routine co-ordination of a Fordist kind; in 'autonomous' flexibly specialised activity). But the debate lacks the basic terms for a critique of power-through-technology in a large sense (for example, anything at the level of 'patriarchy', 'hegemony' or 'Fordism'). At present there are few signs that much attention is being given to the question of which designers are allowed to work on which objects with which users; perhaps some of the research at Aarhus, Denmark—a common locus for Greenbaum and Bødker—is heading this way; perhaps also recent work by Jonathan Grudin (Grudin 1991) at Irvine, California. The human-centred systems movement was about power from the start, even if it could not speak it very clearly, because it originated in a confrontation between capital and labour over material ownership of currently disposed resources—industrial struggles about capital investment, waged work and established skills. In contrast, CSCW is still some way from being pushed to this kind of political edge.

Another level of difficulty—the second 'minus'—is in the absence of significant connections between CSCW and international struggles around new technology. The location of CSCW at the working edge of post- or neo-Fordist global restructuring places it, in principle, in a particularly significant relation with feminist critiques of 'women in the integrated circuit' (Haraway 1985) and with actual transnational struggles of women to recognise and develop the contradictory potentials of new technologies in paid work. Even though CSCW is more debate than technology at present, it is clear that new developmental principles and practical roles in systems development are being explored; within a few years actual commercial technologies will certainly be exploited by organisations operating on a significant national and international scale. It is puzzling, then, that there seems to be little practical link-up between the radicals in the CSCW community and communities of workers struggling internationally around new technology in paid work. One reason might be that feminist discourses about information and communications technologies are often fixed simply on 'women' as an empiricist category; this is true, for example, of much of the business transacted at IFIP international conferences on women and computing. There is too great a distance between some of the most radical and powerful feminist and socialist–feminist thinking and some of the most significantly situated and active feminist struggle around technology.

What should we be doing to develop the gender politics of CSCW? During the next five years or so, CSCW systems will become more robust and (following upon marketing evaluations) more visible in the commercial environment. Dozens of groupware products are already in the market and multimedia work group support systems are in prototype use. Already they show a full range of users being targeted, from the most routine (whose activities are seen as suitable for neo-Fordist co-ordination software on the assembly-line model; systems of this kind are sometimes referred to as 'workflow' systems, and interestingly, some of the targeted workers are software engineers) to the most 'autonomous' (whose work worlds are supported with expensive and naturalistically seamless implementations of leading-edge 'embodied virtuality' technology). The visionary emphasis is at the 'autonomous' end, which of course is where the originators of CSCW place themselves; but the market's voice has yet to start adding its full weight to the balance of interpretations of 'co-operative' work.

It is 'horses for courses', and women struggling with new technologies have seen all of this before. If women are 'cheap' workers, they will get cheaper technology aimed at cost-cutting, speed-up and top-down control; if they are among the 'skilled' and 'autonomous' workers there will be more capital investment per capita, aimed at increased flexibility, responsiveness, co-operation and self-conscious enactment of business goals. Many professionals—and especially women—will be on the confused ground where these two definitions meet, and where notions such as 'usability' and 'user-friendliness' may just as easily mean 'policing' and 'foolproofing'. Despite the co-operation rhetoric, CSCW works both sides of this street, and there is not much in CSCW (as technology) that will stack the implementation outcomes one way or the other. Women will have to go through all the same arguments about hidden skill, the value of different workers' work, job and career content as design issues, and end-user involvement in the practice of selecting, designing, configuring and evaluating new systems.

However, there are aspects to CSCW which provide good ground for such struggles, especially in the methodological as distinct from technological dimensions. Participation is well settled in CSCW as a principle of design and development; users as autonomous and creative actors (rather than cheap-but-fallible machine-substitutes or 'noise in the system') is well grounded, theoretically. But women will still have to struggle over which users get attributed this kind of status in the first place.

Feminist activists in trade unions ought to begin to keep their eyes and ears open for groupware and multimedia products and applications, and to monitor where these new kinds of development are being implemented: which industries, which global locations and which occupational sectors—senior managerial work, expensive technical work, non-technical professional work, routine white-collar work. Activists should also try to open more practical spaces in which 'difficult' variations of feminist thinking can be brought into a working relationship with trade union and workplace struggles on the ground of new technologies. Both CSCW and post-modern feminist perspectives are worth the commitment of some resources, even (perhaps especially) in hard times like the present; and both involve some pain in going beyond the obvious.

7

Gender, unions and the new workplace

Realising the promise?

MIRIAM HENRY

SUZANNE FRANZWAY

> Australians are increasingly recognising the need to speed the pace of change in our workplaces. We must increase our knowledge and understanding of how change can be managed so that it involves workforce participants fully and attracts the commitment of both management and employees ... Making Australia a competitive industrialised country must start in the workplace (Workplace Australia 1992).

The glossy brochure advertising this Australian conference on workplace reform resonates with the new promises: management, labour and unions pulling together to achieve the kind of productive culture required for economic regeneration and international competitiveness. A quick scan of the brochure's buzzwords confirms the discourse of consensus: 'total workforce participation'; 'teambuilding and goal setting'; 'multi-skilling'; 'flexible awards'; 'consultative committees'; 'quality assurance'; 'self-managed work teams'; 'a participatively designed workplace'; and so forth. The conference, called Workplace Australia, was explicitly structured around the participation of workplace teams comprising managers, union representatives and employees. A slight example perhaps, but typical of the new 'best practices' rhetoric accompanying the labour market and industrial relations changes wrought by the imperatives of post-industrial capitalism—changes in which, in Australia, the union movement has been pivotally involved.

While much is promised here, the extent to which the social relations of work have been or can be transformed in the new labour market remains problematic and particularly so, we argue, for women. In ways which have been well documented, women have been particularly disadvantaged in the sex segregated and unequal labour markets created through the symbiotic workings of patriarchy and capitalism. Patriarchal assumptions about 'men's work' and 'women's work',[1] together with gender-biased recognition and valuation of skills, have served to maintain a gendered division of labour which, while shifting at the boundaries, has withstood previous waves of technological change, equal pay and equal employment opportunity legislation and attempts to encourage women into 'non-traditional' areas of work (Cockburn 1983; Game & Pringle 1983; Cockburn 1985; Burton 1987; Franzway et al. 1989; Burton 1991, for example). Perhaps the best that can be said is that for women the current conjuncture offers contradictory possibilities: the promise of a break with the traditions formalised in the 'Harvester' judgment; the reality of perhaps just another episode in the ongoing story of 'gender at work'. These tensions are the focus of this chapter which looks at how unions are grappling with the gender dynamics of the labour market and industrial relations changes taking place in Australia. We delineate first the main features of, and union involvement in, the shift from the 'old' to the 'new' labour market. We then explore the potential gains and losses for women in these moves, examining in particular the impact of feminist unionism on strategies for change.[2]

The 'post-modern condition': new discourses, old conflicts?

It has become almost a truism to observe—from a number of theoretically diverse perspectives—that the shift from Fordist to flexible regimes of accumulation characteristic of post-modern, post-industrial society is transforming the relations of production and the meaning of work (for example, Gorz 1982; Jones

1983; Poster 1984; Mathews 1989; Harvey 1989).[3] According to this general argument, within a globalised economy and driven by the imperatives of the information technologies and their associated modes of organisation, Fordist mass production methods based on economies of scale are being replaced by 'just-in-time' flexible specialisation production based on economies of scope. With the unskilled 'human robot' made redundant by automation, the new workplace instead calls for multi-skilled workers able to respond quickly and intelligently to the ever-shifting demands of niche marketing. In the new labour market, the competitive edge is gained through the quality of the workforce rather than the efficient standardisation of production design. Hence the rigid Taylorist division of labour and the 'mental–manual' divide characteristic of old Fordist production methods must give way to more flexible, holistic approaches to work based on worker co-operation, teaming, quality circles and so on. In this 'global capitalist village' (Falk 1992) then, new relationships of work are created, simultaneously interconnected and fragmented, which, it is suggested, fundamentally undermine the class and social divisions of older forms of capitalism.

Are the changes which are occurring fundamental or cosmetic? Are they progressive or reactionary in their effects? Both pessimistic and optimistic readings are possible. Critics argue that the rhetoric of post-Fordism constitutes a form of ideological incorporation which embraces a Utopian vision of co-operative workplace relations masking the exploitative realities of old-fashioned capitalism. They point to the increasing scope for exploitation of workers in a fragmented labour market where class interests are difficult to protect against the power of global, mobile capital. Wheelwright (1992), for example, asserts that world capitalism is 'less susceptible to the checks and balances of the nation-state and national trade unions and movements. Its freedom to expand and contract, to exploit and relocate is greatly enhanced, and its tendency to foster inequality is accentuated, for there is no world government, no world taxing authority, no global minimum wage or welfare state' (p. 74). Burgess and MacDonald (1990) suggest that the 'flexibility imperative' of post-Fordism denotes a shifting balance of power from labour to capital, a weakening of the role of trade unions in the workplace and increased managerial control.

Others, however, see more positive outcomes. For example, while acknowledging the dangers to organised unionism, Piore and Sabel (1984, p. 278) suggest that flexible specialisation production, with its premium on craft skills, offers workers a 'revitalised' productive role. Halal (cited in Harvey 1989, p. 174) celebrates the shift from the 'authoritarian command' and 'profit-centred big business' of old-style capitalism to the 'participative leadership' and 'democratic free enterprise' imperatives of 'high tech' capitalism. In Australia, John Mathews (1989) has been an enthusiastic advocate of the transformative potential of a post-Fordist labour market in terms of workplace relationships, industrial democracy and access to training and career structures, a position reflective of the ACTU's blueprint for change, *Australia Reconstructed* which saw, in the consensus politics of the new labour market, possibilities for radically altered relationships of work and an advanced form of industrial democracy.

Australian responses

Like all OECD countries, Australia has been inextricably drawn further into the international economy with concomitant pressures to engage in a substantial process of economic restructuring and industrial modernisation. In contrast to countries such as Britain, the United States or New Zealand, the restructuring process was initiated by a Labor government in conjunction with a highly proactive union movement (New Zealand's initial restructuring occurred under a Labor government, but with union resistance and exclusion). From the outset, the Hawke Labor government attempted to juggle competing social and economic demands by harnessing its reforming agenda to a strategy of financial deregulation, microeconomic reform and public sector restructuring: equity through and for the market (MacIntyre 1985; MacIntyre 1989; Yeatman 1990; Henry & Taylor forthcoming).[4] This programme was carried out through a corporatist decision-making framework which accommodated Hawke's consensus politics and gave concrete expression to Labor's 'special relationship' with the union movement as well as, in varying degrees, the women's movement and other social movements (Muetzelfeldt & Bates 1992). The pivotal prices and incomes Accord with the unions is a case in point, with the unions trading wages restraint in return for economic growth, in turn providing Labor with the necessary leverage—and electoral longevity—to carry out its macro- and microeconomic reforms.

The politics of the Accord—and of corporatism—have been contested. Its proponents argue that the Accord mechanisms provided the means for broadening the union's narrow labourist concerns with wages and jobs to a more far-reaching and sophisticated social and political agenda aimed at assisting Australia's long-term economic prosperity. Critics argue that the Accord, whatever its potential, represents the capitulation of labour to the overriding interests of capital (Sharp 1988; Ewer et al. 1991). Preston's (1991) feminist critique suggests that, despite some gains for women within tightly targeted areas, the ascendant economic rationalism framing the Accord agreements has resulted in an increasing residualisation of the social wage, of particular importance to women.

What this all highlights is a kind of 'legitimation crisis' for unions' capacity to mediate in the relationship between capital and labour in the current situation. Undoubtedly, for unions the current transformations are potentially dangerous: the ascendance of economic liberalism, marketplace fragmentation and associated changes in the mode of regulation foreshadow an industrial relations setting potentially inimitable to organised unionism. In this context, the Accord and *Australia Reconstructed* can be viewed as part of the politics of survival of the union movement signalling a major shift in union thinking, a decisive break with its craft union legacy and a broadening of its policy agenda. 'Strategic unionism', as proposed by the ACTU, aimed to initiate a process of union restructuring and radical workplace and industrial relations reform based on a strategy of high productivity, high skills and high wages (Guille 1991). The movement's traditional protectionist stance was abandoned in favour of support for a Swedish-style competitive market economy, with a substantial public sector, a strong welfare system, and a consensus rather than conflict approach to industrial relations. In this, *Australia Reconstructed* exemplified a modernising union movement (or leadership) whose key ideas on restructuring dovetailed with Labor's macro- and microeconomic reform strategies and consensus style politics. Out of this have come the workplace and industrial relations changes at the centre of this discussion: union rationalisation, award restructuring and productivity-based enterprise bargaining. Which brings us now to a more specific examination of women's rather ambiguous place in this agenda.

Women, the new workplace and dilemmas for unions

The legacy of craft unionism and a patriarchal wage-fixing system which jointly served to protect men's jobs from women and keep women underpaid remains strong (O'Donnell & Hall 1988; Burton 1991; Poiner & Wills 1991). Australia's labour market remains notoriously and stubbornly sex-segregated, and women's average earnings remain less than men's—though the gap is not as wide as in some comparable countries. The reason for this lies in another legacy—the women's movement and feminist politics which have operated both through the state and the unions to some considerable effect. While there is of course a very long history of struggle here, recent advances owe much to the space afforded feminist politics by Labor governments at both state and federal level (Sawer 1990; Eisenstein 1991). Notable achievements here include the Whitlam government's support for the ACTU's equal pay claim in 1972 and the plethora of state and federal legislation in the areas of equal employment opportunity, affirmative action and sex discrimination. The corporatist relationship between Labor and the ACTU of the Hawke era has also helped to bring a feminist voice to the political and industrial agenda. Hence child-care provision, while still inadequate, has expanded significantly;

accouchement leave is more readily available and taken; parental leave is on the agenda. There is greater provision for permanent part-time work and other flexible work arrangements. Sexual harassment policies are slowly being implemented. Educationally, women—especially middle-class women—have made advances in compulsory and post-compulsory schooling participation in ways that may contribute to better work options.

However, the gains are by no means universal: working-class girls still leave school early, and their job prospects are dismal. The labour market remains sex and income segregated. Most part-time work is not permanent, and working-class and non-English-speaking women constitute the bulk of a precariously paid casual workforce which is largely non-unionised and beyond the realm of equal employment opportunity or affirmative action legislation. Indeed, the EEO strategy is itself the subject of a substantial debate, both in terms of its overall social agenda and, more pragmatically, in terms of its capacity to make any impact on the lives of most working women. So, what promises are held for women in the new workplace?

Arguably, women have much to gain from the demise of craft unionism which served to protect male strongholds. At the level of rhetoric at least, the provision of career pathways, skills recognition and competency-based training offers opportunities for women to break through the barriers of the sex-segregated labour market. The increased scope for part-time work suits many women with families, and the 'double burden' of home and 'work' may be somewhat alleviated by current reforms in the area of 'workers with family responsibilities' such as more flexible leave provisions and more flexible working hours. However, there are a number of problems which, for convenience, are discussed around three areas: the 'flexible' labour market; award restructuring; enterprise bargaining.

The 'flexible' labour market

The 'flexibility imperative' of post-Fordist production highlights a major area of debate around workplace change. 'Flexibility' serves as a linguistic umbrella for a number of different and potentially competing practices, for example financial flexibility (wages), numerical flexibility (hours, workloads or numbers of workers) and functional flexibility (multi-skilling and broad-banding) (Bramble 1988; Lever-Tracey 1988; Ewer et al. 1991). Unions have generally resisted wages flexibility while actively supporting functional flexibility—issues to which we will return in the discussion on award restructuring and enterprise bargaining below. It is the area of numerical flexibility—often associated with casual, part-time and non-traditional working arrangements—which has perhaps posed the most vexed problems for unions, and for women workers. For many women with family responsibilities, flexible working arrangements are a necessity. For unions, flexible working arrangements pose a threat to union organisation and negotiated work practices. On this issue then, there have always been tensions between union policies and the needs of many women workers. The emergent structure of the new labour market is forcing an accommodation of interests here, though whether to the ultimate benefit of women remains questionable in our view, given the accompanying potential for a strengthened core–peripheral division of labour.

Some argue that the new labour market with its expanded opportunities for part-time work may be beneficial for women, reflected in their growing participation in the workforce. 'They seem to be better at it [part-time work] than men and men don't seem to want these sorts of jobs', as one eminent economist, Bob Gregory, said on ABC's 'Lateline' programme (22 July 1992). Others, however, are less sanguine, pointing to the second-class conditions of most part-time work. They distinguish between a relatively small core workforce of multi-skilled workers with career paths and job security, and an expanding peripheral workforce—mostly women—of part-time, casual, short-term contract and temporary workers 'whose deployment is often structured to match changing business needs . . . maximising flexibility while minimising the organisation's commitment to the worker's job security and career development' (Atkinson, cited in Lever-Tracey 1988, p. 212; see also Harvey 1989, p. 150). The implications for women are obvious and dangerous. Women, for example, currently constitute about three-quarters of the part-time labour force, but frequently under conditions not of their choosing. Women constitute the bulk of the outworker labour force, mostly in precarious conditions (Wajcman and Probert 1988; Junor 1990). Weeks (1987) points to the tension here between the demands of capital for a flexible, casualised labour force and the demands of women for permanent part-time work. As indicated above, unions' traditional response to part-time work has been understandably ambivalent given the perceived threats to full-time work and difficulties of union organisation. However, as Weeks suggests, the real issue here is casualisation rather than part-time work *per se* (and, one could add, lingering patriarchal assumptions about women's 'preferred' marginal attachment to full-time work). The permanent part-time work solution could well accommodate both union concerns and the reality of working life for many women, and indeed in more recent years feminist unionism has contributed to a better understanding of, and commitment to, the idea of permanent part-time work (see also Lever-Tracy, 1988).

Whether this solution will withstand the flexibility imperatives of core-periphery organisation is, however, another matter. What seems to be emerging, rather, is a tendency for functional flexibility (the creation of career paths, multi-skilling, broadbanding, etc.) to be accorded to a 'labour aristocracy' of highly paid workers in the core (post-Fordist) labour force (Ewer et al. 1991, p. 40), with the old Fordist techniques being relegated to the peripheral wastelands—into the often paternalistic realm of homes and small workshops, or into third world countries—well beyond the reach of union intervention or award restructuring and training. As Harvey tellingly comments, 'Struggling against capitalist exploitation in the factory is very different from struggling against a father or uncle who organises family labour into a highly disciplined and competitive sweatshop that works to order for multinational capital' (1989, p. 153). While this view might not accommodate the range of outworking conditions now emerging (for example, the middle-class professionals described in the Probert and Wajcman study, 1988), his general conclusion remains disturbing:

> The transition to flexible accumulation has in fact been marked by a revolution (by no means progressive) in the role of women in labour markets and labour processes during a period when

the women's movement has fought for both greater awareness and improved conditions for what is now more than 40 per cent of the labour force in many of the advanced capitalist countries (p. 155).

Award restructuring and training

Award restructuring has become a key plank of industrial regeneration, signalling improved work practices, career paths based on skills recognition and training, and the demise of debilitating demarcation disputes. For women, award restructuring offers real possibilities, with the potential to recognise and accredit previously unrecognised skills, provide access to skill-related career paths and to redesign jobs in less sex stereotyped ways (McCreadie 1991). Again, however, the promises are ambiguous. For example, the strategy has emerged essentially out of the male-dominated metal trades, characterised by many rigid skills classifications. There is no necessary translation of the principles to female-dominated areas of work, such as retailing, processing and clerical, that have relatively flat classification structures which have been multi-skilled already. Indeed, extending classification scales and formally unpacking and recognising the existing multi-skilled nature of much women's work is likely, from employers' point of view, to lead to pressures for wage rises without the concomitant trade-offs in fewer demarcation disputes characteristic of men's work—and hence is likely to be resisted. Additionally, women's work does not easily lend itself to the creation of career paths—for example, the bridges from process/machine work to engineering design jobs remain elusive, while in clerical work (which operates across rather than through industries) women are most likely to find career paths by moving jobs (Probert 1992).

Award restructuring and the establishment of career paths both rest 'on the acquisition, recognition and utilisation of skills. If access to training and assessment is not equitable, there is the potential for entrenching rather than eliminating inequities' (Henneken 1991, p. 109). However, the training agenda poses real problems for women. For example, formally structured training characteristic of the male-dominated trades areas affords greater opportunities and pay than unaccredited, in-house training currently undertaken by women. There are also tensions between employer demands for firm specific training and union demands for portable work skills. Ewer et al., for example, point to the Business Council of Australia's hostility towards portable qualifications:

> . . . the problem with craft or award based career paths is that they only make the external labour market operate more efficiently. To put this another way, they encourage employees to leave the firm in order to pursue careers (1991, p. 127, citing the BCA).

Hence women whose careers require job portability are likely to be offered only limited opportunities for training. On this more general point, Probert (1992, 450) points to the problematic nature of a credentials-based strategy given the ideological and practical barriers facing many poorly paid women workers in gaining access to credentials. There is, she suggests, 'a huge gap between a theoretical career structure and women being able to realise this potential. Award restructuring without the most vigorous equal opportunity and affirmative action practices will be hollow indeed'.

Enterprise bargaining and the industrial relations setting

Following the Swedish model, the ACTU attempted, in *Australia Reconstructed,* to find an industrial relations strategy which balanced the need for centralised wage and policy setting mechanisms with more flexible arrangements for dealing with industrial relations at the plant or enterprise level. At the level of wage fixing, the ACTU strongly endorsed the existing centralised system. On the other hand, the ACTU also recognised the need for strong local organisation and workplace mobilisation as part of a broader strategy of industrial democracy:

> both Sweden and Norway . . . have focused upon solving problems at the plant and enterprise level, for it is here that the 'engine room' of industrial democracy processes is located. It is of course

> necessary to have navigators and officers on the bridge with a sense of direction and purpose, but the motive force must come from the work place, fuelled by education services and bunkered by central research services (ACTU/TDC 1987, p. 177).

The importance of workplace mobilisation highlights a central dilemma for unions in the new workplace. The logic of Fordist production provided unions with a captive membership. In the new, more fragmented labour market with its individualistic underpinnings, union membership has to be 'earned' against a backdrop of a declining membership and large increases in traditionally non-unionised casual, part-time, and feminised areas of work. For example, between 1976 and 1988, workforce numbers in Australia increased by more than one million, with a corresponding union membership growth of only 23 000 (Berry & Kitchener 1989, pp. 52–3). With the increasing emphasis now on workplace-based negotiations, union strategies have to be more sophisticated and flexible in ways not possible under the old craft union structure. Hence the 'strategic unionism' strategy of creating a small number of large, well-resourced industry-based unions able to reach out into decentralised workplaces and deal more effectively and 'holistically' with the range of problems confronting workers at the plant level.

However, the decentralisation strategy is two-edged for, within the broader, fragmenting logic of flexible accumulation, the union position on centralised wage-fixing has been squeezed, first to acknowledge the principle of productivity-based enterprise bargaining within a centralised wage-fixing umbrella and more recently to accept the notion of centralised wage-fixing as the safety net for most poorly paid workers only. In a short space of time then, Australia's centralised wage-fixing system has become residualised, with the real struggle occurring over the union's position in enterprise negotiations. If this unravels, so of course does much of 'strategic unionism'. With Labor, the union position remains secure. However, the federal Liberal opposition (at the time of writing) has signalled its intention to make unions a voluntary partner in enterprise negotiations in line with countries such as Britain, the United States and New Zealand. At that point, workplace politics will become fierce and the unions' position much more precarious.

Certainly, for workers with weak bargaining power—and most women fall into this category—such moves are likely to constitute yet another barrier to pay equity. Burgmann (1990), for example, points to the correlations between centralised wage-fixing systems and smallest wages gender gaps (Australia and New Zealand—her article preceded the recent changes in New Zealand) and decentralised wage-fixing systems and worst wages gender gaps (Japan and the United States). For Burgmann, enterprise bargaining is likely to strengthen already industrially strong groups (men) and further weaken the mostly female, weakest (see also O'Donnell & Hall 1988; Sawer 1991).

Drawing these threads together

Is there a progressive potential here and, if so, under what circumstances can it be extracted? What leverage exists for women in the unions to shape a feminist agenda for these reforms which takes account of the deeply gendered nature of work? These questions raise further issues about whether there is any space for union intervention.

While acknowledging the difficulties for nation-states in confronting the power of global capitalism, Wheelwright (1992) suggests that the real battles now are not between socialism and capitalism but between competing laissez-faire (Anglo–US) versus statist (Euro–Asian) versions of capitalism. Australia, with its unwavering commitment to level playing fields, razored public services and deregulated economies has opted for the former and with support from the union leadership. Clearly, however, there are tensions now over policy directions within fractions of Australian capital as well as within the state itself and the union movement in the face of persistently high levels of unemployment, continuing trade and balance of payment problems and sluggish economic growth. A debate over economic orthodoxies is finally emerging, highlighted by increasing numbers of anti-economic rationalist publications (for example, Pusey 1991; Carrol & Manne 1992; Rees et al. 1992; Vintila et al. 1992;) and current attempts by Labor and the unions to distinguish and distance their approaches from Liberal-style economic rationalism. But the failure of Australian policies to date highlights a fundamental problem with *Australia Reconstructed*, based as it was on assumptions of full employment, centralised wage fixing and policy integration drawn from the 'statist' approaches of Sweden, Norway and Austria. Australia's laissez faire approach, on the contrary, has in effect surrendered macro policy management to the markets, hence undermining the state's capacity to intervene effectively to fulfil such assumptions (Ewer et al. 1991). This leaves the union movement in a considerably weakened position, operating 'off the back foot' to intervene at the level of health and safety, work practices, equal opportunity and so on—interventions which, while important, can only be of limited success given the broader context.

With respect to gender and union issues, what is interesting are the assumptions that 'equity' will be looked after simply through the imperatives of well managed technological change. For example, in an extreme and naive version of this, John Mathews (1989, p. 182) suggests that the new labour market constitutes a 'clean slate' out of which entirely new, ungendered, work practices can be forged:

> This is where the question of technological change acquires its strategic value. At the point of change, *where there are no established practices to unravel, and there are no existing disparities between groups of workers*, it is open to unions to pursue their equal opportunity policies . . . [emphasis added].

This is also essentially the view implicit in *Australia Reconstructed* and many of the more recent reports on workplace change such as the Finn Report (AEC Review Committee 1991) and the Carmichael Report (NBEET, ESFC 1992). In these documents there is the obligatory reference to equity and disadvantage, but little appreciation of the deep-seated gender dynamics contributing to labour market segregation—and hence little appreciation of the range and depth of strategies required to 'manage' the transition to a new and 'desegregated' labour market.

In a less naive but similarly rosy tribute to the new labour market, Heather Carmody (1991, p. 107), addressing a conference on women and workplace reform, suggested that the business community was 'progressively dismantling their stereotypes about the workforce'. She pointed to the range of 'family conscious' policies being adopted by some of Australia's largest companies, citing examples such as parental leave, home-based work opportunities, easier access to training and sick leave to care for dependants. By contrast, other speakers at the conference were less optimistic, pointing to women's continuing predominant position in the most marginalised and least unionised of workplaces, entrenched authoritarian management attitudes and lack of interest in training (see, for instance, Lafferty 1991; Walsh 1991). Undoubtedly there are truths in both versions, for what we are witnessing—contrary to Mathews' ahistorical perception—is a dynamic between new forces and existing, deeply unequal, relationships and practices. As Cockburn (1983, p. 216) suggests, '[n]ew technology is a force acting from without on established social relations. But within that set of relations . . . are many tensions. The effect that new technology will have will depend not only, or even mainly, on the force of its impact but on the pattern of tensions within the social structure'.

The current structural changes in the labour market are clearly of sufficient magnitude to create a space for some change and for concerted union action, circumscribed though this may be by the rather dismal global context. The new workplace could potentially exemplify a process of industrial democracy in which workers' skills would be appropriately recognised and valued, putting an end to the destructive and inequitable sex and class-based division of labour. The fairly regular stream of 'best practices' vignettes now appearing on various current affairs programs certainly sustain Mathews' utopian concept of a 'new compact between capital and labour' (1989, p. 154) and, we could add, between capitalism and patriarchy (women seem to do quite well in these exemplary public relations exercises). As indicated earlier, a 'well managed' flexible labour market with permanent part-time work located in the core rather than the deregulated, non-unionised periphery might contribute to better working conditions for women not only in terms of job security, skills recognition and status but also in terms of the potential for men to take up combinations of family and paid work responsibilities in ways which might lessen women's 'double burden'. However, with Ewer et al. we share some scepticism as to how these changes should be interpreted and about the burgeoning human resources management industry. Alternatively, then, the new workplace could represent a return to nineteenth-century conditions of work in which the strong are favoured and the weak are exploited—particularly if unions are squeezed out of this 'new compact' as in New Zealand or the United Kingdom.

There is some cautious optimism around at present about the progressive possibilities for women but, as with past gains and contrary to the simplistic assumptions of the new politics of consensus, these will have to be fought for every inch of the way—in the workplace, in broader social policy arenas and within the union movement itself. As McCreadie (1991, p. 35) concludes:

> There are undeniable risks to women in the restructuring process, but we need to avoid an overly deterministic approach which sees the sexual division of labour as unassailable or which merely reverts to defence of the status quo . . . The restructuring of gender relations is central to any strategy for progressive social change. If unions are to be agents in such change they will need to revolutionise the methods and scope of their work. This will mean officials ceding some of their power to their members. In many cases this implies men ceding power to women.

In practical terms, this might be easier said than done.

Changing the agenda: the sexual politics of trade unions

Many activists and commentators argue that unions must adapt to the needs of workers in new workplaces, mainly young workers and women, or else the union movement itself will decline into social and political irrelevance (Berry & Kitchener, 1989; Doran 1989; Shute 1990; Ellem 1992). However, union reaction continues to be determined by the economically powerful unions which are also largely the male-dominated unions. These unions are threatened by changes to the workplace which are eroding the power base of the union movement. Similarly, even though there has been an enormous increase in women's participation in the new workplaces, this has not automatically given women enhanced leverage, either as workers or unionists. Rather, it has become evident that union women must devise strategies in relation to both the demands of capital and to the versions of unionism developed by their male comrades. Award restructuring is perhaps only the most obvious example of the complex issues which women must address.

During the last two decades women have made considerable gains in the Australian trade union movement. Membership share is significant; women occupy leadership positions and 'women's issues' have gained a certain currency in the industrial arena (for example, sexual harassment, maternity leave). But women occupy a curious position in relation to the troubles now afflicting the union movement. While activists and union writers identify women workers as of considerable importance to the union movement's survival, trade unions continue to overlook women, both as workers and as unionists. At most union meetings held at Trades Hall, it will commonly be men on the platform and speaking from the floor; similarly, men frequently write the publications, drawing on male experiences and concerns.

Assumptions about gender underpin this notable absence of women. Since the early 1970s, the concept of 'gender' has been contrasted with that of 'sex', to emphasise the socially constructed basis of sexually differentiated behaviour, and to avoid an essentialism unhelpful to theory (Pringle 1992). Although the concept of gender refers to a system of relations between two sexes, it has in practice been used in the analysis of women's position. As Judy Wajcman has argued (in Chapter 2) this allows people to think that they do not need to deal with gender issues if they are not researching women. The result is that the other sex, men, are seen as sex-neutral, without gender.

This has serious consequences for the theories and practices of trade unionism. It is crucial to recognise that in trade unions, as in all social institutions, 'gender' is about power, conflict, ideologies and change. Thus the recognition of women and of issues relevant to women in trade unions, as elsewhere, involves struggle. What follows from this is the recognition that it is not possible to merely add women on to the agenda of trade unions. Sexual politics in unionism is a critical factor, but it is too often disguised.

During the 1989 ACTU Congress Berry and Kitchener asked *Can Unions Survive?* Spurred by the 1988 ABS statistics which showed a steep decline in union membership, they argued strongly that unions were foolish to ignore women and young workers any longer. Indeed they castigated unions for their current attitudes: '. . . many male union officials still hold prehistoric attitudes towards women, lagging behind even the most conservative groups in society' (Berry & Kitchener 1989, p. 46).

This was a cruel criticism for the movement which claims to uphold the banner for progressive social politics. Their argument was, however, that unless Australian unions responded to the demands of the new workplace, they would become entirely irrelevant. Since women were becoming integral to that workplace they are necessarily part of the solution. In this way Berry and Kitchener's call to 're-organise, restructure and revitalise' gave some impetus to the campaigns which women unionists have been waging for the last twenty years.

The implications of these criticisms remain to be taken seriously. Two major collections on Australian industrial relations published in early 1992 mustered seven women out of a total of fifty-six contributors (Crosby & Easson 1992; Dabscheck et al. 1992). 'Women's issues' are discussed with remarkable brevity and, rather then using gender as an integral analytical concept, discussions on women quickly shift to the characteristics of those industries and occupations which are female dominated. *What Should Unions Do?* (Crosby & Easson 1992) includes useful analyses of the complexities of a rapidly changing union membership in Australia, but little appreciation of the significance of gender. For example, one author claims that the view that '. . . female wages are 30–35 per cent lower than male ones is sheer poppycock' (Clarke 1992, p. 23). Several factors, Clarke suggests, are responsible for this gap, including the fact that men work more hours per week, on average, than women. Leaving aside the assumption that this difference in working hours is not to be considered a legitimate 'cause' of unequal earnings, this kind of statement reveals how rarely the gap between mainstream and feminist research on work and industrial relations is bridged. Feminist activists and researchers on women's wages pointed to the differences in hours, long ago. Furthermore, they have debated why such differences persist (see Ryan & Conlon 1975; O'Donnell & Hall 1988; Sharp & Broomhill 1988; Ryan & Conlon 1989; Burgmann 1990). Very little of this work has been integrated into men's writing on trade unionism in which the sexual politics of work, as well as unionism itself, are persistently overlooked. This raises a number of questions about what is involved politically and theoretically in any strategies aimed at mobilising women in unions to assert their claim on the benefits of technological and organisational change in the workplace.

Footholds

Women have long participated in the paid workforce, and been members of trade unions, albeit at lower rates than men. Despite the recent decline in union members as a proportion of the paid workforce, however, women unionists are now in stronger, more visible positions that ever before. In Australia women constitute almost 42 per cent (ABS March 1992) of the paid workforce. Of these women 35 per cent are unionised compared to 45 per cent of male workers (Pocock 1992, p. 1). While there has been a decline in union membership overall, the number of women members has been increasing. However, this rate of unionisation does not match the increase in workforce participation rates; in the last decade, the number of women joining the workforce increased by 38 per cent compared with an increase of only 16 per cent in unionisation (Shute 1990, p. 3). Nevertheless, women make up the majority of new recruits (72 per cent from 1974 to 1984), if only because male rates of unionisation have fallen so dramatically. Between 1982 and 1990 the number of women unionists increased from 860 700 to 975 800 while the number of male unionists declined from 1 706 900 to 1 683 800 (ABS 1990).

Women have never enjoyed proportional representation in the leadership and decision-making positions in trade unions so that the gains they have made in the last decade have been remarkable. When the South Australian United Trades and Labour Council (UTLC) celebrated its centenary in 1984, only one woman (Elizabeth Johnston) had ever been on its executive. By 1992 five women, including one holding an affirmative action position, made up a quarter of the UTLC executive (compared to 14 per cent on the Executive of the Victorian Trades Hall Council).

Two recent surveys have given much more detail on the representation of women in union positions than was previously available. In 1991 the Victorian Trades Hall supported a survey of unions in Victoria (Nightingale 1991); in 1992 Barbara Pocock and Jane Clarke undertook a similar survey in South Australia (Pocock 1992). Each study surveyed the share of positions held by women and men at every level in the unions, from workplace representatives to union executives; the extent of policy and other initiatives specifically concerned with women; and the effectiveness of union actions in support of women. Both observed that there was some skew in their figures, on the basis that unions with less equitable results were less likely to respond. The South Australian survey achieved a higher response rate (84 per cent of unions, compared to 60 per cent of unions in Victoria covering 80 per cent of total membership) but both were significant, and produced similar results.

Martina Nightingale found that overall, women in Victoria held 23 per cent (25 per cent in South Australia) of all available union official positions (elected and appointed) which, when compared with the overall proportion of union members, left a 'gender gap' of 17 per cent. A larger gap (27 per cent) appeared in relation to the position of secretary, the position which is the most senior of the elected full-time officials and so carries the greatest weight and status in the union movement (Nightingale 1991, p. 38). In South Australia, four women were secretaries (7 per cent). Female-dominated unions are more often than not led by a male secretary, as has been the case throughout the history of the union movement.

In contrast, the gender ratio of shop stewards (workplace representatives) to members is much more equitable. In South Australia the gender gap was only 6.5 per cent but the nature of the job and its status is worth noting: 'the position of shop steward has perhaps the lowest status, and is often the most stressful in the union movement, requiring hard work, considerable personal skills and a commitment to the welfare of fellow workers' (Pocock 1992, p. 14). However, women shop stewards were distributed unevenly as larger unions with a sizeable female membership produced strong results which outweighed poor results in ten out of twelve major industry groups. Since the shop steward position constitutes the first rung on the ladder into more powerful union positions, this situation underlined the seriousness of the problem of representation for women, and suggested that there are few grounds for optimism in the traditionally male-dominated industry sectors.

Each of the Australian states has its own idiosyncratic union history but we may assume that the outline presented here of the relative positions of women and men will be similar across the country. Nor does it differ markedly from most other countries. Women occupy 6.7 per cent of 'top management union positions' in Sweden and 16.7 per cent of these positions in Italy (Kaplan 1992, p. 45). Women are gaining a foothold, but what this obscures is the deeply held and very gendered view that 'real unionists' are male. This affects trade union theory as well as trade union practice.

But not gender

It is astonishing how rarely women merit attention in industrial relations studies. Gender and sexual politics are dealt with by default. The 'unionist' is treated as a gender-neutral being which results in the invisibility not only of women *per se* but of gender. Thus in his 1991 study of shop stewards John Benson says: '. . . as only five out of 346 shop stewards at the time of the survey were female, gender was excluded as a predictor of stewards' role definition' (Benson p. 43). Even where women are present, the specificity of gender differences is still studiously ignored in much of the literature (see, for example, Batstone et al. 1977; Waters 1982; Mathews 1989). Where attention is paid to gender, the analyses tend to focus on the peculiarities of women, especially their location in the segmented labour market. Gender relations again slides out of view as the discussion examines the particular characteristics of the (female-dominated) industry or occupation. Male-dominated industries and male workers constitute the norm, to which most mainstream texts hastily return after a brief excursion around 'women's' issues.

By contrast, feminist research on women workers and trade unionism emphasises the social conditions of women, the male dominance of trade unions and the complexities of gender relations. The touchstone of feminist explanations of the gender gap is the sexual division of labour, at work and at home. A considerable debate focuses on the relation between women's work and unionisation and their family responsibilities for housework and caring work. The recruitment and participation of women is framed by the demands of the workplace and the home; union strategies aimed at improving women's position are constrained by the perceived, non-normative conditions of their workplaces (a difficult enough task), and accommodate the limits imposed on women's participation by family demands. This implies an acceptance of unequal gender divisions within the family. Even initiatives for new union recruitment campaigns and provision for child care at meetings can reflect a desire to avoid the potential conflicts of sexual politics at home and in the workplace and the union.

While women's responsibilities within the domestic division of labour are significant factors, there are equally important constraints which stem from women's occupational and industrial segregation. Much closer attention should be given to the ways in which the limited distribution of women through the workforce reduces the opportunity for active union participation and leadership, including their working hours, the relative concentration of women in small firms, their skills and so on. These constraints are further exaggerated by women's concentration in those occupations and industries, particularly in the service sector, which have less 'industrial muscle'.

Assumptions that women can be 'added on' to the existing trade union agenda fail to account for the powerful complexity of the unequal relations between women and men. This can be seen when an explanation for women's low participation which addresses conflict between women and men is rejected in preference for an analysis which focuses on the relationship between officials and rank-and-file activists (see Ellem's (1989) very interesting account of the clothing trades union). If the arguments were expressed in terms of gender, issues of male identity would be uncovered, thus allowing the attitudes, including hostility, of male unionists to women in union positions to be exposed.

Many feminists have argued that women's relative absence from trade unions is a result of male unionists' actions to exclude them. For some the sexism of trade unionism is to be explained, in part, by the way in which capitalist development undermined men's pride in their masculinity. For Marilyn Lake 'the labour movement arose out of men's experience of work and met men's needs; it was a response to men's degradation as men as well as their exploitation as workers . . . [it arose] from a crisis of masculinity induced by industrialisation' (Lake 1986, pp. 137–9). In Mike Donaldson's work (1991), gender relations in trade unionism are also seen not simply as a reflection of male power, but as a response to particular capitalist strategies to maximise profits. Historically, the treatment of women as second-class workers and unionists meant that men were frequently obliged to defend themselves against the cheap female labour which employers were (and still are) keen to use. The contradictory logic of this position is illustrated in the early demands by trade unions for equal pay: often cynical in intent and designed as a defence against women getting the jobs, it was assumed that no employer would choose a woman if she cost as much as a man (Ryan & Conlon 1989, p. 64). In this context, women's antipathy and relatively low unionisation may have been a reasonable response to male dominance of trade unions and their cursory treatment of women workers (Charles 1983).

The legacy of this history of trade union practice is that unionists are still men unless otherwise specified. It is male unionists who are the focus of historical accounts of trade unionism, studies of union practices or of industrial relations, while female unionists are always women first. As women, female unionists are stereotyped as lacking in militancy and the other qualities associated with good unionism—as being passive, flighty and wayward, and exhibiting extremes of attitude from outright hostility to the greatest of commitment. This widely held view of women as unreliable or apathetic remains in spite of the challenge to it by the record of women's activism. For example, Claire Williams's (1988) study of flight attendants found that they challenged all the accepted criteria of the 'real' unionist, being young women working in small and (literally) mobile workplaces. Yet they were often very militant.

Bea Campbell has called trade unionism a male movement, well aware that trade unions are also organisations of the working class (Campbell 1984). Much of the politics and research on unions has as its premise a commitment to a left-wing politics which assumes the importance of a mass, class base or organisation. Unionism is understood in structural terms, or as a force of (Marxist) history, and so as a manifestation of the real working class. That class is identified with the male gender while the complexities of the interaction of class and gender arise only where the focus is on women. Yet the call for class solidarity was and is as potent for women as for men; women as wage earners have needed and demanded unionisation as evidenced by the English match girls who surprised their leader, Annie Besant, with their strike in 1888 (Boston 1980). The same story can be seen in the militant actions taken by Australian nurses in the 1980s. The intricate difficulties of managing the competing demands of class and gender which this entails for women activists is persuasively illustrated in Meredith Tax's fictionalised account of working women in New York, *Rivington Street.*

Sexual politics are at stake here. Change does not occur through the exercise of reason or good will alone. Changing the agenda involves a sexual politics which points to the issue of power in unions and the question: what kind of power do trade union women have?

Power, leadership and gender

Leaders may be remote, egalitarian or manipulative but they are apparently gender neutral. This is why feminists suggest that new definitions of leadership and power are necessary for the promotion of feminist political change. Nancy Hartsock proposes that power is better understood as 'energy and competence rather than dominance' (Hartsock 1983, p. 224). The problem is that energy and competence do not guarantee a place on the union election ticket, neither is it evident that, having won the election, women will refrain from exercising dominance.

Bob Connell recommends that we consider gender power in terms of a 'structure of power, a set of social relations with some scope and permanence' which he connects with ideologies of male supremacy. Organisational control depends on the ability to assert hegemony which is essential to social power. When authority is defined as legitimate, in the structure of power, authority is connected with masculinity although complicated by the denial of authority to some groups of men. 'Men can enjoy patriarchal power, but accept it as if it were given to them by an external force, by nature or convention or even women themselves, rather than by an active social subordination of women going on here and now' (Connell 1987, p. 215).

Women who seek power face the opposite circumstance: neither nature, convention nor men readily grant women power as an unquestioned given. It is no surprise, therefore, that few women have gained formal authority in trade unions. It is also apparent that the unequal participation of women in unions cannot be explained simply in terms of the nature of female-dominated industries, or addressed through uncertain attempts to 'add on' a few women to the union executive.

A further complication is that union leadership involves a perpetual tension between conflicting goals which act as a constraint on each other:

> [Union leadership] is constantly caught between attempting to provide comprehensive representation for all the interests of its working class constituency and being limited in its ability to find a formula that reconciles these partly contradictory interests without endangering their internal acceptability and/or external negotiability (Offe & Wiesenthal, cited in Gardner 1986, p. 173).

The ways in which these tensions are resolved involves the selective articulation of the interests of members, and thus raises the question of whether women, who are not 'real unionists', can be representative of those interests.

The Nightingale and Pocock surveys show that women nonetheless occupy a significant number of positions in trade unions. When we consider what is involved in taking on positions of union power, as full-time officials or through different levels of activism, other pressures become as critical as masculine hegemony, pressures which are part of actually 'doing the job'. In general, union jobs are highly demanding, with very long hours; Heery and Kelly (1989) found the average to be in excess of 60 hours per week. Huw Beynon describes the incursions of union demands on the shop steward and his family's time including, for example, Sunday morning meetings. In addition, in a telling phrase '. . . there's the drinking on top of that' (1973, p. 203).

Diane Watson's study of union officials confirmed the problem of overwork but admitted that 'we might have to allow that some officers found such intense activity and involvement a satisfying reward and chose to do it' (Watson 1988, p. 122). Women officials make such choices but under rather more complicated conditions. Where all women in the paid workforce have to deal with the demands of family and domestic responsibilities, women unionists find that these can be such obstacles to involvement that they tend to delay their activism until these demands have reduced. The account by Mary Douglas (1992) of her experience of having three babies in three years while she continued to work as a union official is rare for any working woman but almost unique within the union movement. Neither are the difficulties confined to women as mothers. As one young official remarked: 'My leisure time has declined and it's difficult to invest time in important relationships. I'm not married but you would have to be extremely well organised to combine this job with children. It's not impossible but it's certainly not something I'd contemplate at the moment' (Watson 1988, p. 131).

Women officials are not only different from their male counterparts, they are also different from many other women, most of whom have two children, usually before they are thirty. Although their work places great demands on the rest of their lives, trade union men are not faced with the same demands. Heery and Kelly's British study of full time female officials found 54 per cent with no children as did Ledwith et al. (1990). In an as yet unpublished brief survey of twenty-four middle ranking officials who attended a national residential training forum in 1992, Franzway found only three who shared households with children. Of eight high-ranking officials interviewed as part of the same project, six had none, one had grown-up children and one had a child under 5. The last is widely regarded as being extraordinary.

When women gain union positions they must battle further limiting attitudes if they are to succeed as leaders. Roby & Uttal's (1988) study of shop stewards in the US found that women were not passive position holders and demonstrated that relative to men they were highly committed, as well as active in negotiating the appropriate conditions, including family support and resources. The problems have to do not only with winning the numbers but with a sexual politics where women's presence, let alone authority, is continually contested.

Conclusion

The way in which women in significant union positions exercise their formal power is shaped by gender relations, among other things. They contest the structure of power of trade unions as well as challenge and create gender identities and gender relations. This is much more than just 'adding women on'. Rather there is a need for a different vision of leadership and power which would recognise the existence of the private as well as the public domain, and explicitly address the complexity of gender relations.

The enormous difficulties and dilemmas of the current conjuncture have led to some recognition of the significance of women workers and of gender relations. However, there are considerable tensions between feminist unionism and the still dominant trade union agenda and its attempts to develop appropriate strategies in relation to the exigencies of the new workplace. As women have gained a firmer foothold in union structures, they have challenged those structures, and union policies, to change substantially in order that women's needs and perspectives might be incorporated.

If unions decline, women as well as men workers will suffer from the lack of an organised defence. Gender must be integrated into the theories and practices of trade unionism as an essential strategy in the support of workers as well as in the protection of unions. The possibilities offered by new technologies in the workplace will not be realised for women without a different unionism which recognises the importance of gender relations.

Notes

[1] The quotation marks signify the problematic nature of how men's and women's work has come to be defined. However, for the sake of readability, they are omitted in future references.

[2] The union movement is obviously diverse and not necessarily united over policy directions, strategies etc. However, for our purposes terms such as 'unions' or 'the union movement' are used to signify organised unionism as epitomised by the peak body, The Australian Council of Trade Unions (ACTU).

[3] No attempt is made here to explore the theoretical minefield surrounding post-modernism, post-industrialism and post-Fordism. The terms are used here to denote the cultural, technological and labour market domains respectively. See Rose 1991.

[4] While these writers point to the explicit reconceptualisation of equity as a form of economic productivity under Labor since 1983, they also indicate how this process fits into a long tradition of Australian labourism in which the market has been seen as the fundamental provider of welfare.

Bibliography

Acker, Joan (1990), 'Hierarchies, jobs, bodies: a theory of gendered organizations', *Gender and Society*, 4, 2, pp. 139–58.

Albin, Peter and Eileen Appelbaum (1988), 'The computer rationalization of work: implications for women workers', in J. Jenson, E. Hagen, and C. Reddy (eds), *Feminization of the Labor Force: Paradoxes and Promises*, London, Polity Press, pp. 137–52.

Appelbaum, Eileen (1985), Computer rationalization of work: the choice between algorithmic and sociotechnical organizational designs, presented at the Department of Commerce Conference on Human Factors, Productivity and Technology, Washington, DC, September.

—— (1987), 'Technology and the redesign of work in the insurance industry', in B. Wright (ed.), *Women, Work and Technology: Transformations*, Ann Arbor, University of Michigan Press, pp. 182–201.

—— and Peter Albin (1989), 'Computer rationalization and the transformation of work: lessons from the insurance industry', in S. Wood (ed.), *The Transformation of Work?* London, Unwin Hyman, pp. 247–65.

Arnold, E. et al. (1982), 'Microelectronics and Women's Employment in Britain', SPRU Occasional Paper Series No. 17, University of Sussex.

Attewell, Paul (1990), 'Information technology and the productivity paradox', New York, Department of Sociology, Graduate Centre, CUNY.

Australian Bureau of Statistics (August 1990), Catalogue No. 6325.0 *Trade Union Members Australia*, Canberra.

—— (March 1992), Cat. No. 6203.0, *The Labour Force Australia*, Canberra.

Australian Council of Trade Unions/Trade Development Commission (1987), *Australia Reconstructed: A Report by the Mission Members to the ACTU and the TDC*, Canberra, AGPS.

Australian Education Council Review Committee (1991), *Young People's Participation In Post-Compulsory Education And Training* (Finn Report), Canberra, AGPS.

Badham, Richard (1991), 'Technology, work and culture', Guest editorial, *AI and Society*, 5, pp. 261–76.

Bamber, G. J. and R. D. Lansbury (eds) (1989), *New Technology: International Perspectives on Human Resources and Industrial Relations*, London, Unwin Hyman.

Bannon, Liam, Mike Robinson and Kjeld Schmidt (eds) (1991), *ECSCW '91—Proceedings of the second European conference on computer supported cooperative work*, Amsterdam, Kluwer.

Baran, Barbara and Jana Gold (1988), 'New markets and new technologies: work reorganization and changing skill patterns in three white collar service industries', Berkeley, Berkeley Roundtable on the International Economy, University of California at Berkeley.

Baran, Barbara and Carol Parsons (1986), 'Technology and skill: a literature review', Berkeley, Berkeley Roundtable on the International Economy, University of California at Berkeley.

Barker, J. and H. Downing (1980), 'Word processing and the transformation of the patriarchal relations of control in the office', *Capital and Class*, 10, Spring, pp. 64—97.

Barnes, B. and D. Edge (eds) (1982), *Science in Context: Readings in the Sociology of Science*, Milton Keynes, Open University Press.

Batstone, E., I. Boraston and S. Frenkel (1977), *Shop Stewards in Action*, Oxford, Basil Blackwell.

Beale, J. (1982), *Getting It Together: Women as Trade Unionists*, London, Pluto Press.

Beechey, V. (1987), *Unequal Work*, London, Verso.

Belsey, C. (1980), *Critical Practice*, London, Methuen.

Bennet, M. K. (1972), *Secretary: an Enquiry into the Female Ghetto*, London, Sidgwick and Jackson.

Benson, J. (1991), *Unions at the Workplace. Shop Steward Leadership and Ideology*, Melbourne, Oxford University Press.

Berry, P. and Kitchener, G. (1989) *Can Unions Survive?* BWIU, ACT Branch, Canberra.

Bertrand, Olivier and Thierry Noyelle (1988), *Human Resources and Corporate Strategy: Technological Change in Banks and Insurance Companies in Five OECD Countries*, Paris, OECD.

Bijker, W., T. Hughes and T. Pince (eds) (1987), *The Social Construction of Technological Systems,* Cambridge, Mass., MIT Press.

Bjerknes, Gro and Tone Bratteteig (1987), *Florence in Wonderland—Systems Development with Nurses,* in Gro Bjerknes, Pelle Ehn and Morten Kyng (eds), *Computers and Democracy—A Scandinavian Challenge,* Aldershot, Avebury, pp. 279–94.

Bjerknes, Gro, Tone Bratteteig and Karlheinz Kautz (eds) (1992), *Proceedings of the IRIS-15 Conference,* Oslo, Dept of Informatics, University of Oslo.

Bødker, Susanne and Joan Greenbaum (1988), *A non-trivial pursuit—systems development as cooperation; a report of the ROSA project,* Arhus, Denmark, Arhus University Computer Science Department, DAIMI PB-268.

Bødker, Susanne (1991), 'Activity theory as a challenge to systems design', in H. E. Nissen, H. K. Klein and R. Hirschheim (eds) (1991), *Information Systems Research—Contemporary Approaches and Emergent Traditions,* Amsterdam, Elsevier/North-Holland, pp. 551–64.

Boston, S. (1980), *Women Workers and the Trade Union Movement,* London, Davis-Poynter.

Bourdieu, Pierre (1980), *Outline of a Theory of Practice,* Cambridge, Cambridge University Press.

Bramble, T. (1988), 'The flexibility debate: industrial relations and new management production practices', *Labour and Industry,* 1, 2, pp. 187–209.

Burgess, J. and D. Macdonald (1990), 'The labour flexibility imperative', *Journal of Australian Political Economy,* number 27, pp. 15–35.

Burgmann, M. (1990), 'A mistaken enterprise', *Australian Left Review,* No. 19, pp. 15–16.

Burton, C. (with H. Raven and G. Thompson) (1987), *Women's Worth: Pay Equity and Job Evaluation in Australia,* Canberra, AGPS.

Burton, C. (1991), *The Promise and the Price: The Struggle for Equal Opportunity in Women's Employment,* Sydney, Allen and Unwin.

Byrkjeflot, Haldor and Sissel Myklebust (1991), *Technological Change and Human Resources in the Norwegian Service Sector,* Norway, Centre for Technology and Culture, TMV Report Series No. 2.

CACM (1991), Communications of the Association for Computing Machinery, Special Issue on CSCW, December 1991.

Cameron, D. (1985), *Feminism and Linguistic Theory,* London, Macmillan.

Campbell, B. (1984), *Wigan Pier Revisted,* London, Virago.

Carmody, H. (1991), 'Productivity and Equity' in *Balancing the Gains:*

Women, Efficiency and Award Restructuring, Women's Policy Unit, Office of the Cabinet, Queensland, pp. 103–8.

Carnevale, Anthony P. and Harold Goldstein (1990), 'Schooling and training for work in America: an overview', in L. A. Ferman, M. Hoyman, J. Cutcher-Gershenfeld, and E. J. Savoie (eds), *New Developments in Worker Training: A Legacy for the 1990s*, Madison, WI, Industrial Relations Research Association Series.

Carrol, J. and M. Manne (1992), *Shutdown: The Failure of Economic Rationalism and how to Rescue Australia*, Melbourne, Text Publications.

Charles, N. (1983), 'Women and Trade Unions', *Feminist Review*, No. 15. pp. 3–22.

Clark, D. (1992), 'How to save Australia's trade unions from extinction: a three-point radical reform programme', in M. Crosby and M. Easson (eds) (1992), *What Should Unions Do?*, Sydney, Pluto Press, pp. 20–9.

Clement, Andrew (1991), 'Designing without designers–more hidden skill in office computerization?', in Inger V. Eriksson, Barbara A. Kitchenham and Kea G. Tijdens (eds) (1991), *Women, Work and Computerisation—Understanding and Overcoming Bias in Work and Education*, Proceedings of IFIP TC9/WG9.1 Conference, Helsinki, June-July 1991, Amsterdam, Elsevier/North-Holland, pp. 15–32.

Cockburn, Cynthia (1977), *The Local State*, London, Pluto Press.

—— (1983a), 'Caught in the wheels', *Marxism Today*, November.

—— (1983b), *Brothers: Male Dominance and Technological Change*, London, Pluto Press.

—— (1985), *Machinery of Dominance: Men, Women and Technical Know-How*, London, Pluto Press.

Collective Design/Projects (eds) (1985), *Very Nice Work if You Can Get It—The Socially Useful Production Debate*, Nottingham, Spokesman Books.

Connell, R. W. (1987), *Gender and Power*, Sydney, Allen and Unwin.

Cooley, Mike (1980), *Architect or Bee? The Human/Technology Relationship*, Slough, Langley Technical Services.

—— et al. (1989), *European Competitiveness in the 21st Century*, European Commission, FAST, June.

Cowan, Ruth Schwarz (1979), 'From Virginia Dare to Virginia Slims: women and technology in American life', *Technology and Culture*, 20, 1, pp. 51–63.

Cox, E. and H. Leonard (1991), *From Ummm . . . to Aha! Recognising Women's Skills*, Canberra, AGPS, Women's Research and Employment Initiatives Program, Department of Employment, Education and Training.

Crosby, M. and Easson, M. (eds) (1982), *What Should Unions Do?* Sydney, Pluto Press.

Dabscheck, B., G. Griffin and J. Teicher (eds) (1992), *Contemporary Australian Industrial Relations*, Melbourne, Longman Cheshire.

Davis, E. M. and R. D. Lansbury (1989), 'Worker participation in decisions on technological change in Australia', in G. J. Bamber and R. D. Lansbury (eds) *New Technology: International Perspectives on Human Resources and Industrial Relations*, London, Unwin Hyman, pp. 100–16.

de Lauretis, T. (1987), *Technologies of Gender: Essays of Theory, Film and Fiction*, Bloomington, Indiana University Press.

Delgado, A. (1979), *The Enormous File: A Social History of the Office*, London, John Murray.

Dertouzos, Michael L., Richard K. Lester, and Robert M. Solow (1989), *Made In America: Regaining the Productive Edge*, Cambridge, Mass., MIT Press.

Donaldson, M. (1992), *Time of our Lives: Labour and Love in The Working Class*, Sydney, Allen and Unwin.

Doran, J. (1989), 'Unions and Women' in B. Ford and D. Plowman (eds), *Australian Unions. An Industrial Relations Perspective*, 2nd ed. Melbourne, Macmillan, pp. 190–202.

Douglas, M. (1982), 'Women—the essential difference in a reformed trade union movement: one woman's view' in M. Crosby and M. Easson (eds), *What Should Unions Do?*, Sydney, Pluto Press, pp. 362–9.

Easlea, B. (1981), *Science and Sexual Oppression: Patriarchy's Confrontation with Woman and Nature*, London, Weidenfeld and Nicolson.

Ehn, Pelle (1988), *Work Oriented Design of Computer Artefacts*, Stockholm, Arbetslivcentrum.

Ehrenreich, Barbara and John Ehrenreich (1977), 'The professional-managerial class', *Radical America*, 11 (2), reprinted in Pat Walker (ed.) (1979), *Between Labor and Capital*, Hassocks, Harvester Press, pp. 5–45.

Ehrenreich, Barbara and Deirdre English (1979), *For Her Own Good—150 Years of the Experts' Advice to Women*, London, Pluto Press.

Eisenstein, H. (1991), *Gender Shock: Practising Feminism On Two Continents*, Sydney, Allen and Unwin.

Ellem, B. (1989), *In Women's Hands? A History of Clothing Trade Unionism in Australia*, Kensington, University of New South Wales Press.

—— (1992), 'Organising strategies for the 1990's: targeting particular groups: women, migrants, youth' in M. Crosby and M. Easson (eds), *What Should Unions Do?*, Sydney, Pluto Press, pp. 347–61.

Eriksson, Inger V., Barbara A. Kitchenham and Kea G. Tijdens (eds) (1991), *Women, Work and Computerisation—Understanding and Overcoming Bias in Work and Education*, Proceedings of IFIP TC9/WG9.1 Conference, Helsinki, June-July 1991, Amsterdam, Elsevier/North-Holland.

Ewer, P., I. Hampson, C. Lloyd, J. Rainford, S. Rix, and M. Smith, (1991), *Politics and the Accord*, Sydney, Pluto Press.

Falck, Margrit (1991), The interface or the 'golden mean' between human beings at work and technology, paper for IFIP WG9.1 Conference, Human Jobs and Computer Interfaces, Tampere, Finland, June 1991.

Falk, R. (1992), Challenges of a changing global order, paper presented at the International Peace Research Association Fourteenth Annual Conference, Japan.

Fee, E. (1981), 'Women's Nature and Scientific Objectivity' in M. Loew and R. Hubbard (eds), *Woman's Nature: Rationalizations of Inequality*, New York, Pergamon Press, pp. 9–27.

Feldberg, Roslyn L. and Evelyn Nakano Glenn (1979), 'Male and female: job versus gender models in the sociology of work', *Social Problems*, 26 (5), pp. 524–38.

Field, Paul (1985), 'Making people powerful—Coventry workshop', in Collective Design/Projects (eds), *Very Nice Work if You Can Get It—The Socially Useful Production Debate*, Nottingham, Spokeman Books, pp. 53–60.

Finegold, David, and David Soskice (1988), 'The failure of training in Britain: analysis and prescription', *Oxford Review of Economic Policy*, 4 (3), pp. 21–50.

Fleck, J. (1988), *Innofusion or Diffusation? The Nature of Technological Development in Robotics*, Edinburgh PICT Working Paper No. 4.

Flynn, Patrice A. (1990), Does employer-provided training pay off?, paper presented at the Graduate Seminar of the University of Texas at Austin, Austin, October 26.

Franzway, S., D. Court and R. Connell (1989), *Staking a Claim: Feminism, Bureaucracy and the State*, Sydney, Allen and Unwin.

Frizzell, J. (1991), *Identifying Skills in Seven Easy Steps*, Sydney, Pluto Press.

Game, A. and R. Pringle (1983), *Gender at Work*, Sydney, Allen and Unwin.

Gantt, Michelle and Bonnie Nardi (1992), Gardeners and gurus—patterns of cooperation among CAD users, *Proceedings of CHI '92*, May 1992, Monterey, California, pp. 107–17.

Gardner, M. (1986), 'The "fateful meridian": trade union strategies and women workers', in M. Bray and V. Taylor (eds), *Managing Labour? Essays in the Political Economy of Australian Industrial Relations,* Sydney, McGraw-Hill, pp. 168–94.

Gill, Karamjit S. (1990), *Summary of human-centred systems research in Europe,* England, SEAKE Centre, Brighton Polytechnic, on behalf of The Research Institute of System Science, NTT Data, Japan.

Gorz, A. (1982), *A Farewell to the Working Class,* Pluto, London.

Green, Eileen, Jenny Owen and Den Pain (1991), 'Developing computerised office systems—a gender perspective in UK approaches', in Inger V. Eriksson, Barbara A. Kitchenham and Kea G. Tijdens (eds), *Women, Work and Computerisation—Understanding and Overcoming Bias in Work and Education,* Proceedings of IFIP TC9/WG9.1 Conference, Helsinki, June-July 1991, Amsterdam, Elsevier/North-Holland, pp. 217–32.

Green, Eileen, Jenny Owen and Den Pain (eds) (1993), *Gender, Information Technology and Office Systems Design,* London, Falmer Press.

Greenbaum, Joan (1990), 'The head and the heart—using gender analysis to study the social construction of computer systems', *Computers and Society,* 20, 2, June 1990, pp. 9–17.

—— and Morten Kyng (eds) (1991), *Design at Work—Cooperative Design of Computer Systems,* Hillsdale, NJ, Lawrence Erlbaum Associates.

Greenbaum, Joan (forthcoming), 'A design of one's own—towards participatory design in the US', in A. Namioka and D. Schuler (eds), *Participatory Design,* Hillsdale, New Jersey, Lawrence Erlbaum Associates.

Grudin, Jonathan (forthcoming), 'Obstacles to participatory design in large product development organisation', in A. Namioka and D. Schuler (eds), *Participatory Design,* Hillsdale, New Jersey, Lawrence Erlbaum Associates.

Guille, H. (1991), 'Unions and Economics', *Social Alternatives,* 10 (1), pp. 40–3.

Gustafsson, G. (1986), 'Co-determination and wage earner funds', in Fry, J. (ed.) (1986) *Towards a Democratic Rationality,* Aldershot, Gower.

Hacker, Sally L. (1979), 'Sex stratification, technology and organizational change: a longitudinal case study of AT&T', *Social Problems,* 26 (5), pp. 539–57.

—— (1989), *Pleasure, Power and Technology—Some Tales of Gender, Engineering and the Cooperative Workplace,* London, Unwin Hyman.

Hales, Mike (1980), *Living Thinkwork—Where Do Labour Processes Come From?* London, Free Association Books/CSE Books.

—— (1982), *Science or Society? The Politics of the Work of Scientists,* London, Free Association Books.

—— (1988), *Women: The Key to Information Technology.* Briefing Pack for London Strategic Policy Unit.

—— (1991), 'A human-resource approach to information systems development—The ISU (information systems use) design model', *Journal of Information Technology*, 6, pp. 140–61.

—— (1993), 'User participation in design—what it can do, what it can't, and what this means for management', in Paul Quintas (ed.), *The Social Dimensions of Information Systems Engineering*, London, Ellis Horwood.

—— (forthcoming), 'Where are designers? Styles of design practice, objects of design and views of users in computer supported co-operative work', in Duska Rosenberg (ed.), *Design Issues in CSCW*, London, Springer-Verlag.

—— and Peter O'Hara (1993), 'Strengths and weaknesses of participation—learning by doing in local government', in Eileen Green, Jenny Owen and Den Pain (eds), *Gender, Information Technology and Office Systems Design*, London, Falmer Press.

Håpnes, T. and B. Rasmussen (1991), 'The production of male power in computer science', in *Proceedings of the conference on Women, Work and Computerization*, A. M. Lehto and I. Eriksson (eds), Helsinki, Finland, pp. 407–23.

Haraway, Donna (1985), 'A manifesto for cyborgs—science, technology and socialist feminism in the 1980s', *Socialist Review*, 80.

—— (1991), *Simians, Cyborgs and Women—The Reinvention of Nature*, London, Free Association Books.

Harding, S. (1986), *The Science Question in Feminism*, New York, Cornell University Press.

Hart, M. (1984), 'How the Office of the Future is Shaping Up', *Chartered Accountant Magazine*, Vol. 117, August.

Hartmann, Heidi and Roberta Spalter-Roth (1989), 'Women in telecommunications: an exception to the rule', Washington, DC, Institute for Women's Policy Research.

Hartsock, N. (1983), 'The feminist standpoint: developing the ground for a specifically feminist historical materialism', in S. Harding and M. Hintikka (eds), *Discovering Reality: Feminist Perspectives on Epistemology, Metaphysics, Methodology and Philosophy of Science*, Dordrecht, Reidel, pp. 283–310.

—— (1983), *Money Sex and Power, Toward a Feminist Historical Materialism*, New York, Longman.

Harvey, D. (1989), *The Condition Of Postmodernity: An Enquiry Into The Origins Of Cultural Change*, Oxford, Basil Blackwell.

Heery, E. and J. Kelly (1989), ' "A cracking job for a woman'—a profile of women trade union officers', *Industrial Relations Journal*, 20 (3), pp. 192–202.

Henneken, P. (1991), 'Equity issues and the changing training agenda' in *Balancing The Gains: Women, Efficiency and Award Restructuring*, Women's Policy Unit, Office of the Cabinet, Queensland, pp. 109–18.

Henriques, J., W. Holloway, C. Urwin, C. Venn and V. Walkerdine (1984), *Changing the Subject: Psychology, Social Regulation and Subjectivity*, London, Methuen.

Huggett, C. (1988), *Participation in Practice: A Case Study of the Introduction of New Technology*, Watford, Engineering Industry Training Board.

Hughes, T. (1983), *Networks of Power: Electrification in Western Society*, Baltimore and London, John Hopkins University Press.

Hyman, R. (1975), *Industrial Relations: A Marxist Introduction*, London, Macmillan.

—— and W. Streeck (eds) (1988), *New Technology and Industrial Relations*, Oxford, Basil Blackwell.

Illich, Ivan (1975), *Tools for Conviviality*, London, Fontana.

Jenson, J. (1989), 'The talents of women, the skills of men: flexible specialization and women', in S. Wood (ed.), *The Transformation of Work?*, London, Unwin Hyman, pp. 141–55.

Jones, B. (1983), *Sleepers, Wake! Technology and The Future Of Work*, Melbourne, Oxford University Press.

Jordanova, L. J. (1980), 'Natural facts: a historical perspective on science and sexuality', in C. MacCormack and M. Strathern (eds), *Nature, Culture and Gender*, Cambridge, Cambridge University Press.

Junor, A. (1990), *Employment Estimates and Projections Relevant to Commerce and Business Studies, Australia and New Zealand*, unpublished report prepared for Projects of National Significance: Broadening Girls' Post-School Options, Sydney, Macquarie University.

Kaplan, G. (1982), *Contemporary Western European Feminism*, Sydney, Allen and Unwin.

Keller, E. Fox (1983), *A Feeling for the Organism: The Life and Work of Barbara McClintock*, San Francisco, Freeman.

—— (1985), *Reflections on Gender and Science*, New Haven, Yale University Press.

Kelley, Maryellen R. (1989), 'Alternative forms of work organization under programmable automation', in S. Wood (ed.), *The Transformation of Work?* London, Unwin Hyman, pp. 235–46.

Kern, H. and M. Schumann (1987), 'Limits of the division of labour: new production and employment concepts in West German Industry', *Economic and Industrial Democracy*, 8, pp. 151–70.

Kessler-Harris, A. (1985), 'Problems of coalition-building: women and trade unions in the 1920's', in R. Milkman (ed.), *Women, Work and*

Protest: A Century of U.S. Women's Labor History, Boston, Routledge and Kegan Paul, pp. 110–138.

Kochan, Thomas A. and Paul Osterman (1991), 'Human resource development and utilization: is there too little in the U.S.?' Cambridge, Mass., Sloan School of Management, Massachusetts Institute of Technology.

Kuhn, T. (1970), *The Structure of Scientific Revolutions*, Chicago, Chicago University Press.

Kuutti, Kari (1991), 'Activity theory and its applications to information systems research and development', in H. E. Nissen, H. K. Klein and R. Hirschheim (eds), *Information Systems Research—Contemporary Approaches and Emergent Traditions*, Amsterdam, Elsevier/North Holland, pp. 529–49.

—— and Tuula Arvonen (1992), 'Identifying potential CSCW applications by means of activity theory concepts—a case example', in Gro Bjerknes, Tone Bratteteig and Karlheinz Kautz (eds) (1992), *Proceedings of the IRIS-15 Conference*, Oslo, Dept of Informatics, University of Oslo, pp. 217—31.

Lafferty, G. (1991), 'Equity and Employment in the Queensland Tourism and Hospitality Industry' in *Balancing The Gains: Women, Efficiency and Award Restructuring*, Women's Policy Unit, Office of the Cabinet, Queensland, pp. 162–81.

Lake, M. (1986), 'A question of time', in D. McKnight (ed.), *Moving Left: The Future of Socialism in Australia*, Sydney, Pluto Press, pp. 135–48.

Latour, Bruno (1991), 'Technology is society made durable', in John Law (ed.), *A Sociology of Monsters—Essays on Power, Technology and Domination*, London, Routledge, pp. 103–31.

Ledwith, S., et al. (1990), 'The making of women trade union leaders', *Industrial Relations Journal*, 21 (2), pp. 112–25.

Lehto, A. M., and I. Eriksson (eds) (1991), *Proceedings of the Conference on Women, Work and Computerization*, Helsinki.

Lever-Tracy, C. (1988), 'The flexibility debate: part time work', *Labour and Industry*, 1 (2), pp. 210–41.

Liff, S. (1990), 'Clerical workers and information technology: gender relations and occupational change', *New Technology, Work and Employment*, 5 (1), Spring, pp. 44–55.

Lynch, Lisa M. (1989), 'Private sector training and its impact on the wages of young workers', National Bureau of Economic Research Working Paper No. 2872, Cambridge, Mass., NBER.

—— and Paul Osterman (1989), 'Technological innovation and employment in telecommunications', *Industrial Relations*, 28 (2), Spring, pp. 188–205.

Lyytinen, Kalle (1990), 'Computer supported cooperative work—issues and challenges, a structurational analysis', Working paper, Department of Computer Science, University of Jyvâs Kylâ, Finland.

—— and Ojelanki Ngwenyama (1992), 'What does computer support for cooperative work mean? A structurational analysis of CSCW', *Accounting, Management and Information Technology*, 2, 2.

McCreadie, S. (1991), 'Restructuring Australian manufacturing—opportunities for women', *Social Alternatives*, 10 (1), pp. 32–6.

MacIntyre, S. (1985), *Winners and Losers: The Pursuit of Social Justice in Australian History*, Sydney, Allen and Unwin.

—— (1988–89), 'Less Winners, More Losers', *Australian Society Social Justice Supplement*, pp. 35–7.

MacKenzie, D. and J. Wajcman (eds) (1985), *The Social Shaping of Technology*, Milton Keynes, Open University Press.

McNeil, M. (ed.) (1987), *Gender and Expertise*, London, Free Association Books.

Malz, D., and R. A. Borker (1982), 'A cultural approach to male/female miscommunication', in *Language and Social Identity*, ed. John J. Gumperz, Cambridge, Cambridge University Press.

Martin, E. (1991), 'The egg and the sperm: how science has constructed a romance based on stereotypical male-female roles', *Signs*, 16 (31), pp. 485–501.

Martin, R. (1980), *Trade Unions in Australia*, Melbourne, Penguin.

Mathews, J. (1989), *Tools of Change: New Technology and the Democratisation of Work*, Sydney, Pluto Press.

May, F. T. (1981), 'IBM word processing developments', *IBM Journal of Research Development*, 25 (5), September.

Merchant, C. (1980), *The Death of Nature: Women, Ecology and the Scientific Revolution*, New York, Harper and Row.

Merchiers, Jacques (1991), 'Changing skills in metalworking industries: a review of research', *Training and Employment: French Dimension*, Paris, CEREQ Newsletter, No. 4, Summer.

Milkman, R. (ed.) (1985), *Women, Work and Protest. A Century of U.S. Women's Labour History*, Boston, Routledge and Kegan Paul.

Moran, T. P. and R. J. Anderson (1990), 'The workaday world as a paradigm for CSCW design' in *Proceedings of CSCW '90*.

Muetzelfeldt, M. and R. Bates (1992), 'Conflict, contradiction and crisis' in M. Muetzelfeldt (ed.), *Society, State and Politics in Australia*, Sydney, Pluto Press, pp. 43–75.

Murray, F. (1989), 'Beyond the negation of gender: technology bargaining and equal opportunities', *Economic and Industrial Democracy*, 10, pp. 517–42.

Namioka, A. and D. Schuler (eds) (forthcoming), *Participatory Design*, Hillsdale, New Jersey, Lawrence Erlbaum Associates.

Nardi, B. and J. R. Miller (1991), 'Twinkling lights and nested loops—Distributed problem solving and spreadsheet development', *International Journal of Man-Machine Studies*, 34, pp. 161–84.

National Board of Employment, Education and Training, Employment and Skills Formation Council (1992), *Australian Vocational Certificate Training System* (Carmichael Report), Canberra, AGPS.

Nicholson, Linda J. (ed.) (1990), *Feminism/Postmodernism*, London, Routledge.

Nightingale, M. (1991), *Facing the Challenge. Women in Victorian Unions*, Victorian Trades Hall Council, Melbourne.

Noble, D. (1979), 'Social choice in machine design: the case of automatically controlled machine tools', in A. Zimbalist (ed.) *Case Studies on the Labor Process*, New York, Monthly Review Press, pp. 18–50.

Norman, Donald A. and Stephen W. Draper (eds) (1986), *User Centred System Design*, Hillsdale, New Jersey, Lawrence Erlbaum Associates.

O'Donnell, C. and P. Hall (1988), *Getting Equal: Labour Market Regulation and Women's Work*, Sydney, Allen and Unwin.

Office of Technology Assessment (OTA), Congress of the United States (1990), *Worker Training: Competing in the New International Economy*, Washington, DC, U.S. Government Printing Office.

Organisation of Economic Cooperation and Development (1988), *New Technologies in the 1990s: A Socio-economic Approach*, Paris.

Pacey, A. (1983), *The Culture of Technology*, Oxford, Basil Blackwell.

Palmer, L. (1988), 'Telephone exchange maintenance', in E. Willis (ed.), *Technology and the Labour Process*, Sydney, Allen and Unwin, pp. 155–68.

Penn, R. (1982), 'Skilled manual workers in the labour process, 1856—1964' in S. Wood (ed.), *The Degradation of Work? Skill, Deskilling and the Labour Process*, London, Hutchinson, pp. 90–108.

Phillips, A. (1983), 'Review of *Brothers*', *Feminist Review*, No. 15, pp. 101–4.

—— and B. Taylor (1980), 'Sex and skill: notes towards feminist economics', *Feminist Review*, No. 6, pp. 79–88.

Piore, M. and C. Sabel (1984), *The Second Industrial Divide: Possibilities For Prosperity*, New York, Basic Books.

Pocock, B. (1988), *Demanding Skill: Women and Technical Education in Australia*, Sydney, Allen and Unwin.

—— (1992), *Women Count. Women in South Australian Unions*, Centre for Labour Studies and the South Australian United Trades and Labor Council.

Poiner, G. and S. Wills (1991), *The Gifthorse: A Critical Look at Equal Employment Opportunity in Australia*, Sydney, Allen and Unwin.

Poster, M. (1984), *Foucault, Marxism and History: Mode of Production Versus Mode of Information*, Oxford, Polity Press.

Poynton, C. (1985), *Language and Gender: Making the Difference*, Geelong, Deakin University Press.

—— and K. Lazenby (1992), *What's In A Word: Recognition of Women's Skills in Workplace Change*, (Research Report), Adelaide, Women's Department of Industrial Relations.

Preston, B. (1991), 'The accord and the social wage: a lost opportunity?' *Social Alternatives*, 10 (1), pp. 15–18.

Pringle, R. (1988), *Secretaries Talk*, London, Verso.

—— (1992), 'Absolute sex? unpacking the sexuality/gender relationship' in R. W. Connell, and G. W. Dowsett (eds), *Rethinking Sex. Social Theory and Sexuality Research*, Melbourne, Melbourne University Press, pp. 76–101.

Probert, B. (1992), 'Award restructuring and clerical work: skills, training and careers in a feminized occupation', *Journal of Industrial Relations*, 34 (3), pp. 436–54.

Roach, Stephen S. (1988), 'Technology and the services sector: America's hidden competitiveness challenge,' in B. R. Guile and J. B. Quinn (eds), *Technology in Services: Policies for Growth, Trade and Employment*, Washington, DC, National Academy Press.

Roby, P. and L. Uttal (1988), 'Trade union stewards. Handling union, family, and employment responsibilities', *Women and Work: An Annual Review*, 3, pp. 215–48.

Rose, H. (1983), 'Hand, brain, and heart: a feminist epistemology for the natural sciences', *Signs*, 9 (1), pp. 73–90.

Rose, M. (1991), *The Post-Modern and the Post-Industrial: A Critical Analysis*, Cambridge, Cambridge University Press.

Rothschild, J. (ed.) (1983), *Machina Ex Dea: Feminist Perspectives on Technology*, New York, Pergamon Press.

Ryan, E. (1984), *Two-thirds of a Man. Women and Arbitration in New South Wales, 1902–08*, Sydney, Hale and Iremonger.

—— and A. Conlon (1989), *Gentle Invaders: Australian Women at Work*, Melbourne, Penguin.

Sawer, M. (1990), *Sisters in Suits: Women and Public Policy in Australia*, Sydney, Allen and Unwin.

—— (1991), 'Why has the women's movement had more influence on government in Australia than elsewhere?' in F. Castles (ed.), *Australia Compared: People, Policies and Politics*, Sydney, Allen and Unwin, pp. 258–77.

Sayre, A. (1975), *Rosalind Franklin and DNA: A Vivid View of What It Is*

Like to Be a Gifted Woman in an Especially Male Profession, New York, W.W. Norton and Co.

Sharp, G. (1988), 'Reconstructing Australia', *Arena*, 82, pp. 70–88.

Sharp, R. and R. Broomhill (1988), *Short-changed. Women and Economic Policies*, Sydney, Allen and Unwin.

Shute, C. (1990), Don't Be Too Polite, Girls!: Women, Power and the Australian Trade Union Movement, Paper presented to the First World Summit on Women and the Many Dimensions of Power, Montreal, Canada, June.

Silverstone, R. and R. Towler (1983), 'Progression and tradition in the job of secretary', *Personnel Management*, May.

Smith, Stephen (1991), 'On the economic rationale for codetermination law', *Journal of Economic Behavior and Organization*, 16, pp. 261–81.

Sorge, Arndt and Wolfgang Streeck (1988), 'Industrial relations and technical change: the case for an extended perspective', in R. Hyman and W. Streeck (eds), *New Technology and Industrial Relations*, Oxford, Basil Blackwell, pp. 19–47.

Sorge, Arndt and Malcolm Warner (1986), *Comparative Factory Organization: An Anglo-German Comparison of Manufacturing, Management and Manpower*, Aldershot, Gower.

Spender, D. (1980), *Man Made Language*, London, Routledge and Kegan Paul.

Stacey, M. (1960), *Tradition and Change*, Oxford, Oxford University Press.

Stanley, A. (1992), *Mothers and Daughters of Invention: Notes for a Revised History of Technology*, Metuchen, New Jersey, Scarecrow Press.

Star, Susan Leigh (1991), 'Invisible work and silenced dialogues in knowledge representation', in Inger V. Eriksson, Barbara A. Kitchenham and Kea G. Tijdens (eds), *Women, Work and Computerisation—Understanding and Overcoming Bias in Work and Education*, Proceedings of IFIP TC9/WG9.1 Conference, Helsinki, June–July 1991, Amsterdam, Elsevier/North-Holland, pp. 81–92.

Stevenson, Laurie and Barbara Lepani (1991), *Finance Sector Union Strategies in a World of Change*, Wollongong, The Centre for Technology and Social Change, University of Wollongong.

Storper, Michael and Richard Walker (1983), 'The theory of labour and the theory of location', *International Journal of Urban and Regional Research*, 7 (1), pp. 1–43.

Strathern, M. (1980), 'No nature, no culture: the Hagen case' in C. MacCormack and M. Strathern (eds), *Nature, Culture and Gender*, Cambridge, Cambridge University Press, pp. 174–222.

Streeck, Wolfgang (1991), 'On the institutional conditions of diversified quality production', in Em Matzner and W. Streeck (eds.), *Beyond Keynesianism: The Socio-Economics of Production and Full Employment*, London, Edward Elgar, pp. 21–61.

Suchman, Lucy (1987), *Plans and Situated Actions—The Problem of Human-Machine Communication*, Cambridge, Cambridge University Press.

—— (1991), 'Identities and differences', Closing remarks to the Fourth Conference on Women, Work and Computerisation, in Inger V. Eriksson, Barbara A. Kitchenham and Kea G. Tijdens (eds) (1991), *Women, Work and Computerisation—Understanding and Overcoming Bias in Work and Education*, Proceedings of IFIP TC9/WG9.1 Conference, Helsinki, June-July 1991, Amsterdam, Elsevier/North-Holland, pp. 431–7.

Tallard, M. (1988), 'Bargaining over new technology: a comparison of France and West Germany', in R. Hyman and W. Streeck (eds), *New Technology and Industrial Relations*, Oxford, Basil Blackwell, pp. 284–96.

Tannen, D. (1991), *They Just Don't Understand: Women and Men in Conversation*, Sydney, Random House Australia.

Tax, M. (1991), *Rivington Street*, New York, Monthly Review Press.

Thompson, D. (1989), ' "The sex/gender" distinction: a reconsideration', *Australian Feminist Studies*, No. 10, Summer, pp. 23–32.

Thorne, B., C. Kramarae and N. Henley (eds) (1983), *Language, Gender and Society*, Rowley, Mass., Newbury House.

Threadgold, T. (1988), 'Language and gender', *Australian Journal of Feminist Studies*, 6, pp. 41–69.

Tijdens, K. et al. (1989), *Women, Work and Computerisation: Forming New Alliances*, Amsterdam, Elsevier/North Holland.

Toffler, A. (1980), *The Third Wave*, London, Collins/Pan.

Turkle, S. (1984), *The Second Self: Computer and The Human Spirit*, London, Granada.

Vehvilainen, Marja (1986), 'A study circle as a method for women to develop their work and computer systems', Dublin, Second IFIP conference on Women, Work and Computerisation.

—— (1991), 'Gender in information systems development—a women office workers' standpoint', in Inger V. Eriksson, Barbara A. Kitchenham and Kea G. Tijdens (eds) (1991), *Women, Work and Computerisation—Understanding and Overcoming Bias in Work and Education*, Proceedings of IFIP TC9/WG9.1 Conference, Helsinki, June-July 1991, Amsterdam, Elsevier/North-Holland, pp. 247–61.

Vinnicombe, S. (1980), *Secretaries, Management and Organizations*, London, Heinemann.

Vintila, P., J. Phillimore and P. Newman (eds) (1992), *Markets, Morals and Manifestos: Fightback! And the Politics of Economic Rationalism in the 1990s*, Perth, ISTP, Murdoch University.

Wainright, Hilary and Dave Elliott (1982), *The Lucas Plan—A New Trade Unionism in the Making?* London, Allison and Busby.

Wajcman, J. and Probert, B. (1988) 'New technology outwork' in E. Willis (ed.), *Technology And The Labour Process*, Sydney, Allen and Unwin, pp. 51–67.

Wajcman, Judy (1991), *Feminism Confronts Technology*, North Sydney, Allen and Unwin.

Walby, S. (1986), *Patriarchy at Work*, Cambridge, Polity Press.

Walker, Pat (ed.) (1979), *Between Labor and Capital*, Hassocks, Harvester Press.

Walsh, B. (1991), 'Equity restructuring in retail industry—union perspective', in *Balancing The Gains: Women, Efficiency and Award Restructuring*, Women's Policy Unit, Office of the Cabinet, Queensland, pp. 63–75.

Waters, M. (1982), *Strikes in Australia. A Sociological Analysis of Industrial Conflict*, Sydney, Allen and Unwin.

Watson, D. (1988), *Managers of Discontent. Trade Union Officers and Industrial Managers*, London, Routledge and Kegan Paul.

Webster, J. (1990a), *Office Automation: The Labour Process and Women's Work in Britain*, Hemel Hempstead, Harvester Wheatsheaf.

—— (1990b), 'The Shaping of Software Systems in Manufacturing Issues in the Generation and Implementation of Network Technologies in British Industries', Edinburgh PICT Working Paper No. 17.

Weedon, C. (1987), *Feminist Practice and Postructuralist Theory*, Oxford, Basil Blackwell.

Weeks, W. (1987), 'Part time work: paradox in a shrinking labour market', *Australian Journal of Social Issues*, 22 (3), pp. 517–29.

Wheelwright, T. (1992) 'Global capitalism now: depression in the 1990s', *Arena*, 98, pp. 63–75.

White, J. (1986), 'The writing on the wall: beginning or end of a girl's career?', *Women's Studies International Forum* 9 (5), pp. 561–74.

Williams, C. (1988), *Blue, White and Pink Collar Workers in Australia: Technicians, Bank Employees and Flight Attendants*, Sydney, Allen and Unwin.

Windsor, K. (1991), *Skill Counts: How to Conduct Gender-Bias Free Skills Audits*, Canberra, Women's Employment, Education and Training Advisory Group, Department of Employment, Education and Training, AGPS.

Wood, S. (ed.) (1989), *The Transformation of Work?*, London, Unwin Hyman.

Workplace Australia (1992), 'Conference on Workplace Reform' Sheraton Brisbane Hotel and Towers, June-July.

Yeatman, A. (1990), *Bureaucrats, Technocrats, Femocrats: Essays on the Contemporary Australian State*, Sydney, Allen and Unwin.

Zimmerman, J. (ed.) (1983), *The Technological Women: Interfacing with Tomorrow*, New York, Praeger.

Zuboff, Shoshana (1988), *In the Age of the Smart Machine: The Future of Work and Power*, New York, Basic Books.

Index